Yamaha
FJ600, FZ600, XJ600 and YX600 Radian Owners Workshop Manual

**by John G Edwards
and John H Haynes**

Member of the Guild of Motoring Writers

Models covered:

USA: Yamaha FJ600. 598cc. 1984 and 1985
Yamaha FZ600. 598cc. 1986 through 1988
Yamaha YX600 Radian. 598cc. 1986 through 1990
UK: Yamaha XJ600. 598cc. 1984 through 1992
Yamaha FZ600. 598cc. 1987 and 1988

(2100-7N3)

Haynes UK
Sparkford Nr Yeovil
Somerset BA22 7JJ England

Haynes North America, Inc
859 Lawrence Drive
Newbury Park
California 91320 USA
www.haynes.com

Acknowledgements

Our thanks to Mitsui Machinery Sales (UK) Ltd for permission to reproduce certain illustrations used in this manual. We would also like to thank the Avon Rubber Company for supplying information on tire fitting. Special thanks to Simi Valley Cycles, Simi Valley, California, for supplying the motorcycle, and the Ventura County Superintendent of Schools' Regional Occupational Program - Auto Engine Rebuilding class for the loan of special equipment used in some of the photographs.

A book in the Haynes Owners Workshop Manual Series

ISBN-10: 1-56392-100-6

ISBN-13: 978-1-56392-100-1

Library of Congress Catalog Card Number 94-73122

British Library Cataloguing in Publication Data
A catalogue record for this book is available from the British Library

94-216

Contents

Right side view of the Yamaha FZ600

Right side view of the Yamaha YX600 Radian

About this manual

Its purpose

The purpose of this manual is to help you get the best value from your motorcycle. It can do so in several ways. It can help you decide what work must be done, even if you choose to have it done by a dealer service department or a repair shop; it provides information and procedures for routine maintenance and servicing; and it offers diagnostic and repair procedures to follow when trouble occurs.

We hope you use the manual to tackle the work yourself. For many simpler jobs, doing it yourself may be quicker than arranging an appointment to get the vehicle into a shop and making the trips to leave it and pick it up. More importantly, a lot of money can be saved by avoiding the expense the shop must pass on to you to cover its labor and overhead costs. An added benefit is the sense of satisfaction and accomplishment that you feel after doing the job yourself.

Using the manual

The manual is divided into Chapters. Each Chapter is divided into numbered Sections, which are headed in bold type between horizontal lines. Each Section consists of consecutively numbered paragraphs.

At the beginning of each numbered Section you will be referred to any illustrations which apply to the procedures in that Section. The reference numbers used in illustration captions pinpoint the pertinent Section and the Step within that Section. That is, illustration 3.2 means the illustration refers to Section 3 and Step (or paragraph) 2 within that Section.

Procedures, once described in the text, are not normally repeated. When its necessary to refer to another Chapter, the reference will be given as Chapter and Section number. Cross references given without use of the word "Chapter" apply to Sections and/or paragraphs in the same Chapter. For example, "see Section 8Ó means in the same Chapter.

References to the left or right side of the vehicle assume you are sitting on the seat, facing forward.

Motorcycle manufacturers continually make changes to specifications and recommendations, and these, when notified, are incorporated into our manuals at the earliest opportunity.

Even though we have prepared this manual with extreme care, neither the publisher nor the author can accept responsibility for any errors in, or omissions from, the information given.

NOTE

A **Note** provides information necessary to properly complete a procedure or information which will make the procedure easier to understand.

CAUTION

A **Caution** provides a special procedure or special steps which must be taken while completing the procedure where the Caution is found. Not heeding a Caution can result in damage to the assembly being worked on.

WARNING

A **Warning** provides a special procedure or special steps which must be taken while completing the procedure where the Warning is found. Not heeding a Warning can result in personal injury.

Introduction to the Yamaha FJ600, FZ600, XJ600 and YX600 Radian

The Yamaha FJ/XJ600, FZ600 and YX600 Radian are three different types of motorcycles developed around the same engine and transmission. The FJ600 and XJ600 are sport-touring bikes, the FZ600 is a sport bike and the YX600 Radian is a standard motorcycle.

The engine is an air-cooled, inline four with double overhead camshafts and two valves per cylinder. The design has remained essentially the same since its introduction in 1984.

Fuel is delivered through four Mikuni BS30 or BS32 carburetors.

The rear suspension on the YX600 Radian uses a pair of shock absorber/spring units mounted from the frame to the swingarm. The rear suspension on the FJ600, FZ600 and XJ600 models uses the Yamaha Monocross design, which employs a shock absorber/spring unit mounted ahead of the swingarm. The suspension provides a progressive damping effect. Spring preload and shock absorber damping are adjustable.

The front brakes use dual discs and the rear brakes use a single disc (FJ600, FZ600 and XJ600 models) or a drum brake (YX600 Radian model).

Identification numbers

The frame serial number is stamped into the right side of the frame and printed on a label affixed to the frame center brace. The engine number is stamped into the right upper side of the crankcase. Both of these numbers should be recorded and kept in a safe place so they can be furnished to law enforcement officials in the event of a theft.

The frame serial number, engine serial number and carburetor identification number should also be kept in a handy place (such as with your driver's license) so they are always available when purchasing or ordering parts for your machine.

The models covered by this manual are as follows:

FJ600, FZ600, XJ600 and YX600 Radian - 1984 through 1992

Identifying model years

The procedures in this manual identify the bikes by model year. To determine which model year a given machine is, look for the following identification codes in the engine and frame numbers. **Note:** *UK models can be further identified by their initial engine/frame no. given in parentheses after the identification code in the table.*

Model and Year	Code
FJ600	
1984 and 1985	49A (49 states), 51K (California)
FZ600	
1986 and 1987 US	2AX (49 states), 2AY (California)
1988 US	2XL (49 states), 3BW1 (California)
1987 UK	2HW (2HW-000101)
1988 UK	3BX (2HW-002101)
XJ600	
1984 through 1987	51J (51J-000101)
1989	3KM1 (51J-051101)
1990	3KM3 (51J-070101)
1991 and 1992	3KM5 (3KM-000101)
YX600 Radian	
1986 and 1987	1UJ (49 states), 1UL (California)
1988	2WY (49 states), 2XA (California)
1989	3LT1 (49 states), 3LT2 (California)
1990	3LT3 (49 states), 3LT4 (California)

The frame number is stamped in the right side of the steering head

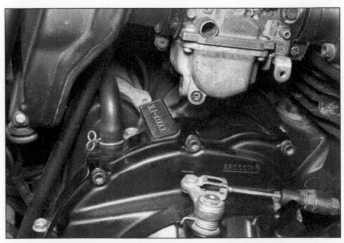

The engine number is stamped into the right side of the crankcase

Buying parts

Once you have found all the identification numbers, record them for reference when buying parts. Since the manufacturers change specifications, parts and vendors (companies that manufacture various components on the machine), providing the ID numbers is the only way to be reasonably sure that you are buying the correct parts.

Whenever possible, take the worn part to the dealer so direct comparison with the new component can be made. Along the trail from the manufacturer to the parts shelf, there are numerous places that the part can end up with the wrong number or be listed incorrectly.

The two places to purchase new parts for your motorcycle - the accessory store and the franchised dealer - differ in the type of parts they carry. While dealers can obtain virtually every part for your motorcycle, the accessory dealer is usually limited to normal high wear items such as shock absorbers, tune-up parts, various engine gaskets, cables, chains, brake parts, etc. Rarely will an accessory outlet have major suspension components, cylinders, transmission gears, or cases.

Used parts can be obtained for roughly half the price of new ones, but you can't always be sure of what you're getting. Once again, take your worn part to the wrecking yard (breaker) for direct comparison.

Whether buying new, used or rebuilt parts, the best course is to deal directly with someone who specializes in parts for your particular make.

General specifications

Wheelbase
  FJ600 .... 1425 mm (56.1 inches)
  FZ600 .... 1385 mm (54.5 inches)
  XJ600 .... 1430 mm (56.3 inches)
  YX600 Radian .... 1385 mm (54 inches)
Overall length
  FJ600 .... 2115 mm (83.3 inches)
  FZ600 (US) .... 2040 mm (80.3 inches)
  FZ600 (UK) .... 2025 mm (79.7 inches)
  XJ600 .... 2145 mm (84.4 inches)
  YX600 Radian .... 2075 mm (81.7 inches)
Overall width
  FJ600 .... 735 mm (28.9 inches)
  FZ600 .... 690 mm 27.2 inches)
  XJ600
    1984 and 1985 .... 735 mm (28.9 inches)
    1989-on .... 745 mm (29.3 inches)
  YX600 Radian .... 770 mm (30.3 inches)
Overall height
  FJ600 .... 1225 mm (48.2 inches)
  FZ600 .... 1145 mm (45.1 inches)
  XJ600 .... 1225 mm (48.2 inches)
  YX600 Radian .... 1095 mm (43.1 inches)
Seat height
  FJ600 .... 790 mm (31.1 inches)
  FZ600 .... 785 mm (30.9 inches)
  XJ600 .... 790 mm (31.1 inches)
  YX600 Radian .... 765 mm (30.1 inches)
Ground clearance (minimum)
  FJ600 .... 140 mm (5.5 inches)
  FZ600 .... 135 mm (5.31 inches)
  XJ600 .... 140 mm (5.51 inches)
  YX600 Radian .... 145 mm (5.7 inches)
Weight (with oil and full fuel tank)
  FJ600 .... 213 kg (470 lbs)
  FZ600 .... 202 kg (409 lbs)
  XJ600 .... 208 kg (459 lbs)
  YX600 Radian .... 197 kg (434 lbs)

Maintenance techniques, tools and working facilities

Basic maintenance techniques

There are a number of techniques involved in maintenance and repair that will be referred to throughout this manual. Application of these techniques will enable the amateur mechanic to be more efficient, better organized and capable of performing the various tasks properly, which will ensure that the repair job is thorough and complete.

Fastening systems

Fasteners, basically, are nuts, bolts and screws used to hold two or more parts together. There are a few things to keep in mind when working with fasteners. Almost all of them use a locking device of some type (either a lock washer, locknut, locking tab or thread adhesive). All threaded fasteners should be clean, straight, have undamaged threads and undamaged corners on the hex head where the wrench fits. Develop the habit of replacing all damaged nuts and bolts with new ones.

Rusted nuts and bolts should be treated with a penetrating oil to ease removal and prevent breakage. Some mechanics use turpentine in a spout type oil can, which works quite well. After applying the rust penetrant, let it work for a few minutes before trying to loosen the nut or bolt. Badly rusted fasteners may have to be chiseled off or removed with a special nut breaker, available at tool stores.

If a bolt or stud breaks off in an assembly, it can be drilled out and removed with a special tool called an E-Z out (or screw extractor). Most dealer service departments and motorcycle repair shops can perform this task, as well as others (such as the repair of threaded holes that have been stripped out).

Flat washers and lock washers, when removed from an assembly, should always be replaced exactly as removed. Replace any damaged washers with new ones. Always use a flat washer between a lock washer and any soft metal surface (such as aluminum), thin sheet metal or plastic. Special locknuts can only be used once or twice before they lose their locking ability and must be replaced.

Tightening sequences and procedures

When threaded fasteners are tightened, they are often tightened to a specific torque value (torque is basically a twisting force). Over-tightening the fastener can weaken it and cause it to break, while under-tightening can cause it to eventually come loose. Each bolt, depending on the material it's made of, the diameter of its shank and the material it is threaded into, has a specific torque value, which is noted in the Specifications. Be sure to follow the torque recommendations closely.

Fasteners laid out in a pattern (i.e. cylinder head bolts, engine case bolts, etc.) must be loosened or tightened in a sequence to avoid warping the component. Initially, the bolts/nuts should go on finger tight only. Next, they should be tightened one full turn each, in a criss-cross or diagonal pattern. After each one has been tightened one full turn, return to the first one tightened and tighten them all one half turn, following the same pattern. Finally, tighten each of them one quarter turn at a time until each fastener has been tightened to the proper torque. To loosen and remove the fasteners the procedure would be reversed.

Disassembly sequence

Component disassembly should be done with care and purpose to help ensure that the parts go back together properly during reassembly. Always keep track of the sequence in which parts are removed. Take note of special characteristics or marks on parts that can be installed more than one way (such as a grooved thrust washer on a shaft). It's a good idea to lay the disassembled parts out on a clean surface in the order that they were removed. It may also be helpful to make sketches or take instant photos of components before removal.

When removing fasteners from a component, keep track of their locations. Sometimes threading a bolt back in a part, or putting the washers and nut back on a stud, can prevent mixups later. If nuts and bolts can't be returned to their original locations, they should be kept in a compartmented box or a series of small boxes. A cupcake or muffin tin is ideal for this purpose, since each cavity can hold the bolts and nuts from a particular area (i.e. engine case bolts, valve cover bolts, engine mount bolts, etc.). A pan of this type is especially helpful when working on assemblies with very small parts (such as the carburetors and the valve train). The cavities can be marked with paint or tape to identify the contents.

Whenever wiring looms, harnesses or connectors are separated, it's a good idea to identify the two halves with numbered pieces of masking tape so they can be easily reconnected.

Gasket sealing surfaces

Throughout any motorcycle, gaskets are used to seal the mating surfaces between components and keep lubricants, fluids, vacuum or pressure contained in an assembly.

Many times these gaskets are coated with a liquid or paste type gasket sealing compound before assembly. Age, heat and pressure can sometimes cause the two parts to stick together so tightly that they are very difficult to separate. In most cases, the part can be loosened by striking it with a soft-faced hammer near the mating surfaces. A regular hammer can be used if a block of wood is placed between the hammer and the part. Do not hammer on cast parts or parts that could be easily damaged. With any particularly stubborn part, always recheck to make sure that every fastener has been removed.

Avoid using a screwdriver or bar to pry apart components, as they can easily mar the gasket sealing surfaces of the parts (which must remain smooth). If prying is absolutely necessary, use a piece of wood, but keep in mind that extra clean-up will be necessary if the wood splinters.

After the parts are separated, the old gasket must be carefully scraped off and the gasket surfaces cleaned. Stubborn gasket material can be soaked with a gasket remover (available in aerosol cans) to soften it so it can be easily scraped off. A scraper can be fashioned from a piece of copper tubing by flattening and sharpening one end. Copper is recommended because it is usually softer than the surfaces to be scraped, which reduces the chance of gouging the part. Some gaskets can be removed with a wire brush, but regardless of the method used, the mating surfaces must be left clean and smooth. If for some reason the gasket surface is gouged, then a gasket sealer thick enough to fill scratches will have to be used during reassembly of the components. For most applications, a non-drying (or semi-drying) gasket sealer is best.

Hose removal tips

Hose removal precautions closely parallel gasket removal precautions. Avoid scratching or gouging the surface that the hose mates against or the connection may leak. Because of various chemical reactions, the rubber in hoses can bond itself to the metal spigot that the hose fits over. To remove a hose, first loosen the hose clamps that secure it to the spigot. Then, with slip joint pliers, grab the hose at the clamp and rotate it around the spigot. Work it back and forth until it is completely free, then pull it off (silicone or other lubricants will ease removal if they can be applied between the hose and the outside of the spigot). Apply the same lubricant to the inside of the hose and the outside of the spigot to simplify installation.

Spark plug gap adjusting tool

Feeler gauge set

Control cable pressure luber

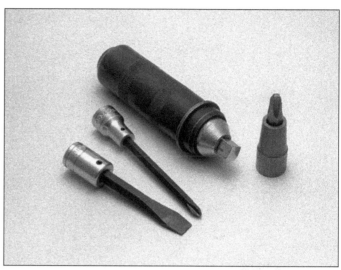

Hand impact screwdriver and bits

If a hose clamp is broken or damaged, do not reuse it. Also, do not reuse hoses that are cracked, split or torn.

Tools

A selection of good tools is a basic requirement for anyone who plans to maintain and repair a motorcycle. For the owner who has few tools, if any, the initial investment might seem high, but when compared to the spiraling costs of routine maintenance and repair, it is a wise one.

To help the owner decide which tools are needed to perform the tasks detailed in this manual, the following tool lists are offered: Maintenance and minor repair, Repair and overhaul and Special. The newcomer to practical mechanics should start off with the Maintenance and minor repair tool kit, which is adequate for the simpler jobs. Then, as confidence and experience grow, the owner can tackle more difficult tasks, buying additional tools as they are needed. Eventually the basic kit will be built into the Repair and overhaul tool set. Over a period of time, the experienced do-it-yourselfer will assemble a tool set complete enough for most repair and overhaul procedures and will add tools from the Special category when it is felt that the expense is justified by the frequency of use.

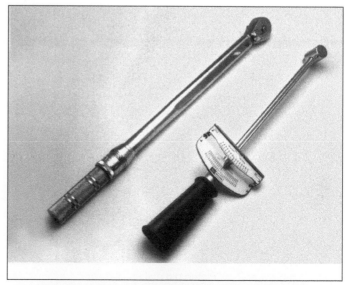

Torque wrenches (left - click type; right - beam type)

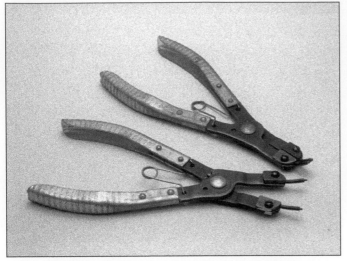

Snap-ring pliers (top - external; bottom - internal)

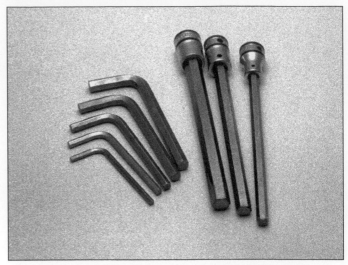

Allen wrenches (left) and Allen head sockets (right)

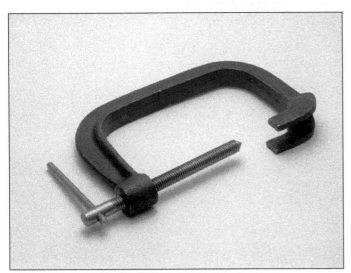

Valve spring compressor

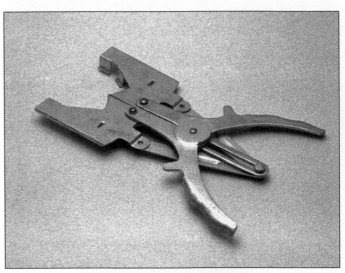

Piston ring removal/installation tool

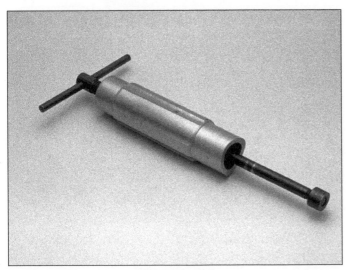

Piston pin puller

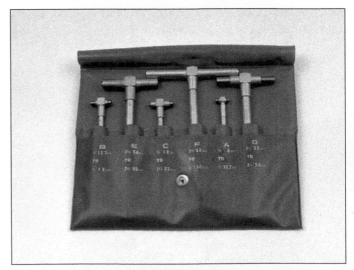

Telescoping gauges

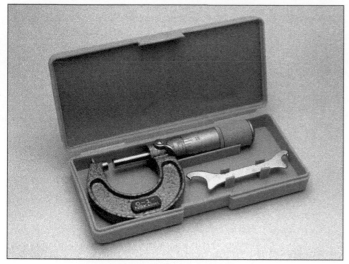

0-to1-inch micrometer

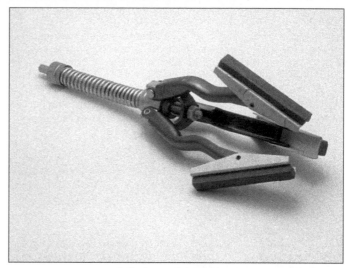

Cylinder surfacing hone

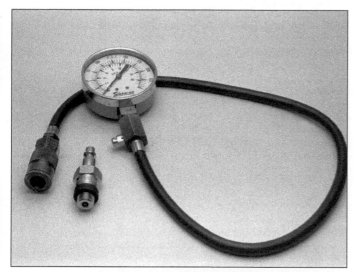

Cylinder compression gauge

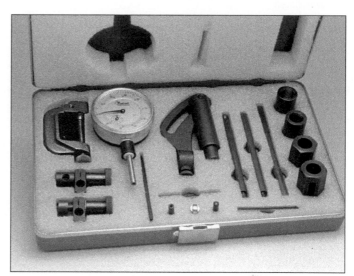

Dial indicator set

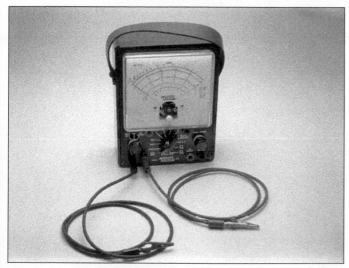

Multimeter (volt/ohm/ammeter)

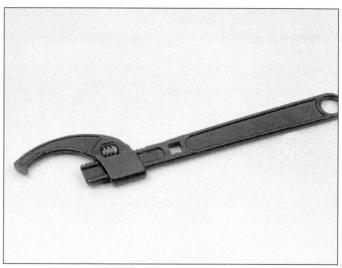

Adjustable spanner

Maintenance and minor repair tool kit

The tools in this list should be considered the minimum required for performance of routine maintenance, servicing and minor repair work. We recommend the purchase of combination wrenches (box end and open end combined in one wrench); while more expensive than open-ended ones, they offer the advantages of both types of wrench.

Combination wrench set (6 mm to 22 mm)
Adjustable wrench - 8 in
Spark plug socket (with rubber insert)
Spark plug gap adjusting tool
Feeler gauge set
Standard screwdriver (5/16 in x 6 in)
Phillips screwdriver (No. 2 x 6 in)
Allen (hex) wrench set (4 mm to 12 mm)
Combination (slip-joint) pliers - 6 in
Hacksaw and assortment of blades
Tire pressure gauge
Control cable pressure luber
Grease gun
Oil can
Fine emery cloth
Wire brush
Hand impact screwdriver and bits
Funnel (medium size)
Safety goggles
Drain pan
Work light with extension cord

Repair and overhaul tool set

These tools are essential for anyone who plans to perform major repairs and are intended to supplement those in the Maintenance and minor repair tool kit. Included is a comprehensive set of sockets which, though expensive, are invaluable because of their versatility (especially when various extensions and drives are available). We recommend the 3/8 inch drive over the 1/2 inch drive for general motorcycle maintenance and repair (ideally, the mechanic would have a 3/8 inch drive set and a 1/2 inch drive set).

Socket set(s)
Reversible ratchet
Extension - 6 in
Universal joint
Torque wrench (same size drive as sockets)
Ball peen hammer - 8 oz
Soft-faced hammer (plastic/rubber)
Standard screwdriver (1/4 in x 6 in)
Standard screwdriver (stubby - 5/16 in)
Phillips screwdriver (No. 3 x 8 in)
Phillips screwdriver (stubby - No. 2)
Pliers - locking
Pliers - lineman's
Pliers - needle nose
Pliers - snap-ring (internal and external)
Cold chisel - 1/2 in
Scriber
Scraper (made from flattened copper tubing)
Center punch
Pin punches (1/16, 1/8, 3/16 in)
Steel rule/straightedge - 12 in
Pin-type spanner wrench
A selection of files
Wire brush (large)

Note: *Another tool which is often useful is an electric drill with a chuck capacity of 3/8 inch (and a set of good quality drill bits).*

Special tools

The tools in this list include those which are not used regularly, are expensive to buy, or which need to be used in accordance with their manufacturer's instructions. Unless these tools will be used frequently, it is not very economical to purchase many of them. A consideration would be to split the cost and use between yourself and a friend or friends (i.e. members of a motorcycle club).

This list primarily contains tools and instruments widely available to the public, as well as some special tools produced by the vehicle manufacturer for distribution to dealer service departments. As a result, references to the manufacturer's special tools are occasionally included in the text of this manual. Generally, an alternative method of doing the job without the special tool is offered. However, sometimes there is no alternative to their use. Where this is the case, and the tool can't be purchased or borrowed, the work should be turned over to the dealer service department or a motorcycle repair shop.

Valve spring compressor
Piston ring removal and installation tool
Piston pin puller
Telescoping gauges
Split-ball gauges
Micrometer(s) and/or dial/Vernier calipers
Cylinder deglazing hone
Cylinder compression gauge
Dial indicator set with clamp and/or magnetic base
Multimeter
Adjustable spanner
Manometer or vacuum gauge set
Small air compressor with blow gun and tire chuck

Buying tools

For the do-it-yourselfer who is just starting to get involved in motorcycle maintenance and repair, there are a number of options available when purchasing tools. If maintenance and minor repair is the extent of the work to be done, the purchase of individual tools is satisfactory. If, on the other hand, extensive work is planned, it would be a good idea to purchase a modest tool set from one of the large retail chain stores. A set can usually be bought at a substantial savings over the individual tool prices (and they often come with a tool box). As additional tools are needed, add-on sets, individual tools and a larger tool box can be purchased to expand the tool selection. Building a tool set gradually allows the cost of the tools to be spread over a longer period of time and gives the mechanic the freedom to choose only those tools that will actually be used.

Tool stores and motorcycle dealers will often be the only source of some of the special tools that are needed, but regardless of where tools are bought, try to avoid cheap ones (especially when buying screwdrivers and sockets) because they won't last very long. There are plenty of tools around at reasonable prices, but always aim to purchase items which meet the relevant national safety standards. The expense involved in replacing cheap tools will eventually be greater than the initial cost of quality tools.

It is obviously not possible to cover the subject of tools fully here. For those who wish to learn more about tools and their use, there is a book entitled *Motorcycle Workshop Practice Manual* (Book no. 1454) available from the publishers of this manual. It also provides an introduction to basic workshop practice which will be of interest to a home mechanic working on any type of motorcycle.

Care and maintenance of tools

Good tools are expensive, so it makes sense to treat them with respect. Keep them clean and in usable condition and store them properly when not in use. Always wipe off any dirt, grease or metal chips before putting them away. Never leave tools lying around in the work area.

Some tools, such as screwdrivers, pliers, wrenches and sockets, can be hung on a panel mounted on the garage or workshop wall, while others should be kept in a tool box or tray. Measuring instruments, gauges, meters, etc. must be carefully stored where they can't be damaged by weather or impact from other tools.

When tools are used with care and stored properly, they will last a very long time. Even with the best of care, tools will wear out if used frequently. When a tool is damaged or worn out, replace it; subsequent jobs will be safer and more enjoyable if you do.

Working facilities

Not to be overlooked when discussing tools is the workshop. If anything more than routine maintenance is to be carried out, some sort of suitable work area is essential.

It is understood, and appreciated, that many home mechanics do not have a good workshop or garage available and end up removing an engine or doing major repairs outside (it is recommended, however, that the overhaul or repair be completed under the cover of a roof).

A clean, flat workbench or table of comfortable working height is an absolute necessity. The workbench should be equipped with a vise that has a jaw opening of at least four inches.

As mentioned previously, some clean, dry storage space is also required for tools, as well as the lubricants, fluids, cleaning solvents, etc. which soon become necessary.

Sometimes waste oil and fluids, drained from the engine or cooling system during normal maintenance or repairs, present a disposal problem. To avoid pouring them on the ground or into a sewage system, simply pour the used fluids into large containers, seal them with caps and take them to an authorized disposal site or service station. Plastic jugs (such as old antifreeze containers) are ideal for this purpose.

Always keep a supply of old newspapers and clean rags available. Old towels are excellent for mopping up spills. Many mechanics use rolls of paper towels for most work because they are readily available and disposable. To help keep the area under the motorcycle clean, a large cardboard box can be cut open and flattened to protect the garage or shop floor.

Whenever working over a painted surface (such as the fuel tank) cover it with an old blanket or bedspread to protect the finish.

Safety first

Professional mechanics are trained in safe working procedures. However enthusiastic you may be about getting on with the job at hand, take the time to ensure that your safety is not put at risk. A moment's lack of attention can result in an accident, as can failure to observe simple precautions.

There will always be new ways of having accidents, and the following is not a comprehensive list of all dangers; it is intended rather to make you aware of the risks and to encourage a safe approach to all work you carry out on your bike.

Essential DOs and DON'Ts

DON'T start the engine without first ascertaining that the transmission is in neutral.

DON'T suddenly remove the filler cap from a hot cooling system - cover it with a cloth and release the pressure gradually first, or you may get scalded by escaping coolant.

DON'T attempt to drain oil until you are sure it has cooled sufficiently to avoid scalding you.

DON'T grasp any part of the engine or exhaust system without first ascertaining that it is cool enough not to burn you.

DON'T allow brake fluid or antifreeze to contact the machine's paint work or plastic components.

DON'T siphon toxic liquids such as fuel, hydraulic fluid or antifreeze by mouth, or allow them to remain on your skin.

DON'T inhale dust - it may be injurious to health (see *Asbestos* heading).

DON'T allow any spilled oil or grease to remain on the floor - wipe it up right away, before someone slips on it.

DON'T use ill fitting wrenches or other tools which may slip and cause injury.

DON'T attempt to lift a heavy component which may be beyond your capability - get assistance.

DON'T rush to finish a job or take unverified short cuts.

DON'T allow children or animals in or around an unattended vehicle.

DON'T inflate a tire to a pressure above the recommended maximum. Apart from over-stressing the carcass and wheel rim, in extreme cases the tire may blow off forcibly.

DO ensure that the machine is supported securely at all times. This is especially important when the machine is blocked up to aid wheel or fork removal.

DO take care when attempting to loosen a stubborn nut or bolt. It is generally better to pull on a wrench, rather than push, so that if you slip, you fall away from the machine rather than onto it.

DO wear eye protection when using power tools such as drill, sander, bench grinder etc.

DO use a barrier cream on your hands prior to undertaking dirty jobs - it will protect your skin from infection as well as making the dirt easier to remove afterwards; but make sure your hands aren't left slippery. Note that long-term contact with used engine oil can be a health hazard.

DO keep loose clothing (cuffs, ties etc. and long hair) well out of the way of moving mechanical parts.

DO remove rings, wristwatch etc., before working on the vehicle - especially the electrical system.

DO keep your work area tidy - it is only too easy to fall over articles left lying around.

DO exercise caution when compressing springs for removal or installation. Ensure that the tension is applied and released in a controlled manner, using suitable tools which preclude the possibility of the spring escaping violently.

DO ensure that any lifting tackle used has a safe working load rating adequate for the job.

DO get someone to check periodically that all is well, when working alone on the vehicle.

DO carry out work in a logical sequence and check that everything is correctly assembled and tightened afterwards.

DO remember that your vehicle's safety affects that of yourself and others. If in doubt on any point, get professional advice.

IF, in spite of following these precautions, you are unfortunate enough to injure yourself, seek medical attention as soon as possible.

Asbestos

Certain friction, insulating, sealing and other products - such as brake pads, clutch linings, gaskets, etc. - contain asbestos. *Extreme care must be taken to avoid inhalation of dust from such products since it is hazardous to health*. If in doubt, assume that they *do* contain asbestos.

Fire

Remember at all times that gasoline (petrol) is highly flammable. Never smoke or have any kind of naked flame around, when working on the vehicle. But the risk does not end there - a spark caused by an electrical short-circuit, by two metal surfaces contacting each other, by careless use of tools, or even by static electricity built up in your body under certain conditions, can ignite gasoline (petrol) vapor, which in a confined space is highly explosive. Never use gasoline (petrol) as a cleaning solvent. Use an approved safety solvent.

Always disconnect the battery ground (earth) terminal before working on any part of the fuel or electrical system, and never risk spilling fuel on to a hot engine or exhaust.

It is recommended that a fire extinguisher of a type suitable for fuel and electrical fires is kept handy in the garage or workplace at all times. Never try to extinguish a fuel or electrical fire with water.

Fumes

Certain fumes are highly toxic and can quickly cause unconsciousness and even death if inhaled to any extent. Gasoline (petrol) vapor comes into this category, as do the vapors from certain solvents such as trichloroethylene. Any draining or pouring of such volatile fluids should be done in a well ventilated area.

When using cleaning fluids and solvents, read the instructions carefully. Never use materials from unmarked containers - they may give off poisonous vapors.

Never run the engine of a motor vehicle in an enclosed space such as a garage. Exhaust fumes contain carbon monoxide which is extremely poisonous; if you need to run the engine, always do so in the open air or at least have the rear of the vehicle outside the workplace.

The battery

Never cause a spark, or allow a naked light near the vehicle's battery. It will normally be giving off a certain amount of hydrogen gas, which is highly explosive.

Always disconnect the battery ground (earth) terminal before working on the fuel or electrical systems (except where noted).

If possible, loosen the filler plugs or cover when charging the battery from an external source. Do not charge at an excessive rate or the battery may burst.

Take care when topping up, cleaning or carrying the battery. The acid electrolyte, even when diluted, is very corrosive and should not be allowed to contact the eyes or skin. Always wear rubber gloves and goggles or a face shield. If you ever need to prepare electrolyte yourself, always add the acid slowly to the water; never add the water to the acid.

Electricity

When using an electric power tool, inspection light etc., always ensure that the appliance is correctly connected to its plug and that, where necessary, it is properly grounded (earthed). Do not use such appliances in damp conditions and, again, beware of creating a spark or applying excessive heat in the vicinity of fuel or fuel vapor. Also ensure that the appliances meet national safety standards.

A severe electric shock can result from touching certain parts of the electrical system, such as the spark plug wires (HT leads), when the engine is running or being cranked, particularly if components are damp or the insulation is defective. Where an electronic ignition system is used, the secondary (HT) voltage is much higher and could prove fatal.

Motorcycle chemicals and lubricants

A number of chemicals and lubricants are available for use in motorcycle maintenance and repair. They include a wide variety of products ranging from cleaning solvents and degreasers to lubricants and protective sprays for rubber, plastic and vinyl.

Contact point/spark plug cleaner is a solvent used to clean oily film and dirt from points, grime from electrical connectors and oil deposits from spark plugs. It is oil free and leaves no residue. It can also be used to remove gum and varnish from carburetor jets and other orifices.

Carburetor cleaner is similar to contact point/spark plug cleaner but it usually has a stronger solvent and may leave a slight oily reside. It is not recommended for cleaning electrical components or connections.

Brake system cleaner is used to remove grease or brake fluid from brake system components (where clean surfaces are absolutely necessary and petroleum-based solvents cannot be used); it also leaves no residue.

Silicone-based lubricants are used to protect rubber parts such as hoses and grommets, and are used as lubricants for hinges and locks.

Multi-purpose grease is an all purpose lubricant used wherever grease is more practical than a liquid lubricant such as oil. Some multi-purpose grease is colored white and specially formulated to be more resistant to water than ordinary grease.

Gear oil (sometimes called gear lube) is a specially designed oil used in transmissions and final drive units, a s well as other areas where high friction, high temperature lubrication is required. It is available in a number of viscosities (weights) for various applications.

Motor oil, of course, is the lubricant specially formulated for use in the engine. It normally contains a wide variety of additives to prevent corrosion and reduce foaming and wear. Motor oil comes in various weights (viscosity ratings) of from 5 to 80. The recommended weight of the oil depends on the seasonal temperature and the demands on the engine. Light oil is used in cold climates and under light load conditions; heavy oil is used in hot climates and where high loads are encountered. Multi-viscosity oils are designed to have characteristics of both light and heavy oils and are available in a number of weights from 5W-20 to 20W-50.

Gas (petrol) additives perform several functions, depending on their chemical makeup. They usually contain solvents that help dissolve gum and varnish that build up on carburetor and intake parts. They also serve to break down carbon deposits that form on the inside surfaces of the combustion chambers. Some additives contain upper cylinder lubricants for valves and piston rings.

Brake fluid is a specially formulated hydraulic fluid that can withstand the heat and pressure encountered in brake systems. Care must be taken that this fluid does not come in contact with painted surfaces or plastics. An opened container should always be resealed to prevent contamination by water or dirt.

Chain lubricants are formulated especially for use on motorcycle final drive chains. A good chain lube should adhere well and have good penetrating qualities to be effective as a lubricant inside the chain and on the side plates, pins and rollers. Most chain lubes are either the foaming type or quick drying type and are usually marketed as sprays.

Degreasers are heavy duty solvents used to remove grease and grime that may accumulate on engine and frame components. They can be sprayed or brushed on and, depending on the type, are rinsed with either water or solvent.

Solvents are used alone or in combination with degreasers to clean parts and assemblies during repair and overhaul. The home mechanic should use only solvents that are non-flammable and that do not produce irritating fumes.

Gasket sealing compounds may be used in conjunction with gaskets, to improve their sealing capabilities, or alone, to seal metal-to-metal joints. Many gasket sealers can withstand extreme heat, some are impervious to gasoline and lubricants, while others are capable of filling and sealing large cavities. Depending on the intended use, gasket sealers either dry hard or stay relatively soft and pliable. They are usually applied by hand, with a brush, or are sprayed on the gasket sealing surfaces.

Thread cement is an adhesive locking compound that prevents threaded fasteners from loosening because of vibration. It is available in a variety of types for different applications.

Moisture dispersants are usually sprays that can be used to dry out electrical components such as the fuse block and wiring connectors. Some types can also be used as treatment for rubber and as a lubricant for hinges, cables and locks.

Waxes and polishes are used to help protect painted and plated surfaces from the weather. Different types of paint may require the use of different types of wax polish. Some polishes utilize a chemical or abrasive cleaner to help remove the top layer of oxidized (dull) paint on older vehicles. In recent years, many non-wax polishes (that contain a wide variety of chemicals such as polymers and silicones) have been introduced. These non-wax polishes are usually easier to apply and last longer than conventional waxes and polishes.

Troubleshooting

Contents

Engine doesn't start or is difficult to start

1 Starter motor does not rotate

1 Engine kill switch Off.
2 Fuse blown. Check fuse block (Chapter 8).
3 Battery voltage low. Check and recharge battery (Chapter 8).
4 Starter motor defective. Make sure the wiring to the starter is secure. Make sure the starter relay clicks when the start button is pushed. If the relay clicks, then the fault is in the wiring or motor.
5 Starter relay faulty. Check it according to the procedure in Chapter 8.
6 Starter button not contacting. The contacts could be wet, corroded or dirty. Disassemble and clean the switch (Chapter 8).
7 Wiring open or shorted. Check all wiring connections and harnesses to make sure that they are dry, tight and not corroded. Also check for broken or frayed wires that can cause a short to ground (see wiring diagrams, Chapter 8).
8 Ignition switch defective. Check the switch according to the procedure in Chapter 8. Replace the switch with a new one if it is defective.
9 Engine kill switch defective. Check for wet, dirty or corroded contacts. Clean or replace the switch as necessary (Chapter 8).
10 Faulty starter lockout switch (if equipped). Check the wiring to the switch and the switch itself according to the procedures in Chapter 8.

2 Starter motor rotates but engine does not turn over

1 Starter motor clutch defective. Inspect and repair or replace (Chapter 2).
2 Damaged idler or starter gears. Inspect and replace the damaged parts (Chapter 2).

3 Starter works but engine won't turn over (seized)

Seized engine caused by one or more internally damaged components. Failure due to wear, abuse or lack of lubrication. Damage can include seized valves, valve lifters, camshaft, pistons, crankshaft, connecting rod bearings, or transmission gears or bearings. Refer to Chapter 2 for engine disassembly.

4 No fuel flow

1 No fuel in tank.
2 Fuel tap vacuum hose broken or disconnected.
3 Tank cap air vent obstructed. Usually caused by dirt or water. Remove it and clean the cap vent hole.
4 Fuel tap clogged. Remove the tap and clean it and the filter (Chapter 1).
5 Fuel line clogged. Pull the fuel line loose and carefully blow through it.
6 Inlet needle valve clogged. For all of the valves to be clogged, either a very bad batch of fuel with an unusual additive has been used, or some other foreign material has entered the tank. Many times after a machine has been stored for many months without running, the fuel turns to a varnish-like liquid and forms deposits on the inlet needle valves and jets. The carburetors should be removed and overhauled if draining the float bowls doesn't solve the problem.

5 Engine flooded

1 Float level too high. Check and adjust as described in Chapter 3.
2 Inlet needle valve worn or stuck open. A piece of dirt, rust or other debris can cause the inlet needle to seat improperly, causing excess fuel to be admitted to the float bowl. In this case, the float chamber should be cleaned and the needle and seat inspected. If the needle and seat are worn, then the leaking will persist and the parts should be replaced with new ones (Chapter 3).
3 Starting technique incorrect. Under normal circumstances (i.e., if all the carburetor functions are sound) the machine should start with little or no throttle. When the engine is cold, the choke should be operated and the engine started without opening the throttle. When the engine is at operating temperature, only a very slight amount of throttle should be necessary. If the engine is flooded, turn the fuel tap off and hold the throttle open while cranking the engine. This will allow additional air to reach the cylinders. Remember to turn the fuel tap back on after the engine starts.

6 No spark or weak spark

1 Ignition switch Off.
2 Engine kill switch turned to the Off position.
3 Battery voltage low. Check and recharge battery as necessary (Chapter 8).
4 Spark plug dirty, defective or worn out. Locate reason for fouled plug(s) using spark plug condition chart and follow the plug maintenance procedures in Chapter 1.
5 Spark plug cap or secondary (HT) wiring faulty. Check condition. Replace either or both components if cracks or deterioration are evident (Chapter 4).
6 Spark plug cap not making good contact. Make sure that the plug cap fits snugly over the plug end.
7 Igniter defective. Check the unit, referring to Chapter 4 for details.
8 Pickup coil(s) defective. Check the unit(s), referring to Chapter 4 for details.
9 Ignition coil(s) defective. Check the coils, referring to Chapter 4.
10 Ignition or kill switch shorted. This is usually caused by water, corrosion, damage or excessive wear. The switches can be disassembled and cleaned with electrical contact cleaner. If cleaning does not help, replace the switches (Chapter 8).
11 Wiring shorted or broken between:
   a) *Ignition switch and engine kill switch (or blown fuse)*
   b) *Igniter and engine kill switch*
   c) *Igniter and ignition coil*
   d) *Ignition coil and plug*
   e) *Igniter and pickup coil(s)*

   Make sure that all wiring connections are clean, dry and tight. Look for chafed and broken wires (Chapters 4 and 8).

7 Compression low

1 Spark plug loose. Remove the plug and inspect the threads. Reinstall and tighten to the specified torque (Chapter 1).
2 Cylinder head not sufficiently tightened down. If the cylinder head is suspected of being loose, then there's a chance that the gasket or head is damaged if the problem has persisted for any length of time. The head nuts should be tightened to the proper torque in the correct sequence (Chapter 2).
3 Improper valve clearance. This means that the valve is not closing completely and compression pressure is leaking past the valve. Check and adjust the valve clearances (Chapter 1).
4 Cylinder and/or piston worn. Excessive wear will cause compression pressure to leak past the rings. This is usually accompanied by worn rings as well. A top end overhaul is necessary (Chapter 2).
5 Piston rings worn, weak, broken, or sticking. Broken or sticking piston rings usually indicate a lubrication or carburetion problem that causes excess carbon deposits or seizures to form on the pistons and rings. Top end overhaul is necessary (Chapter 2).

6   Piston ring-to-groove clearance excessive. This is caused by excessive wear of the piston ring lands. Piston replacement is necessary (Chapter 2).

7   Cylinder head gasket damaged. If the head is allowed to become loose, or if excessive carbon build-up on the piston crown and combustion chamber causes extremely high compression, the head gasket may leak. Retorquing the head is not always sufficient to restore the seal, so gasket replacement is necessary (Chapter 2).

8   Cylinder head warped. This is caused by overheating or improperly tightened head bolts. Machine shop resurfacing or head replacement is necessary (Chapter 2).

9   Valve spring broken or weak. Caused by component failure or wear; the spring(s) must be replaced (Chapter 2).

10   Valve not seating properly. This is caused by a bent valve (from over-revving or improper valve adjustment), burned valve or seat (improper carburetion) or an accumulation of carbon deposits on the seat (from carburetion or lubrication problems). The valves must be cleaned and/or replaced and the seats serviced if possible (Chapter 2).

8   Stalls after starting

1   Improper choke action. Make sure the choke rod is getting a full stroke and staying in the out position.

2   Ignition malfunction. See Chapter 4.

3   Carburetor malfunction. See Chapter 3.

4   Fuel contaminated. The fuel can be contaminated with either dirt or water, or can change chemically if the machine is allowed to sit for several months or more. Drain the tank and float bowls (Chapter 3).

5   Intake air leak. Check for loose carburetor-to-intake manifold connections, loose or missing vacuum gauge access port cap or hose, or loose carburetor top (Chapter 3).

6   Engine idle speed incorrect. Turn throttle stop screw until the engine idles at the specified rpm (Chapters 1 and 3).

9   Rough idle

1   Ignition malfunction. See Chapter 4.

2   Idle speed incorrect. See Chapter 1.

3   Carburetors not synchronized. Adjust carburetors with vacuum gauge or manometer set as described in Chapter 1.

4   Carburetor malfunction. See Chapter 3.

5   Fuel contaminated. The fuel can be contaminated with either dirt or water, or can change chemically if the machine is allowed to sit for several months or more. Drain the tank and float bowls (Chapter 3).

6   Intake air leak. Check for loose carburetor-to-intake manifold connections, loose or missing vacuum gauge access port cap or hose, or loose carburetor top (Chapter 3).

7   Air cleaner clogged. Service or replace air filter element (Chapter 1).

Poor running at low speed

10   Spark weak

1   Battery voltage low. Check and recharge battery (Chapter 8).

2   Spark plug fouled, defective or worn out. Refer to Chapter 1 for spark plug maintenance.

3   Spark plug cap or high tension wiring defective. Refer to Chapters 1 and 5 for details on the ignition system.

4   Spark plug cap not making contact.

5   Incorrect spark plug. Wrong type, heat range or cap configuration. Check and install correct plugs listed in Chapter 1. A cold plug or one with a recessed firing electrode will not operate at low speeds without fouling.

6   Igniter defective. See Chapter 4.

7   Signal generator defective. See Chapter 4.

8   Ignition coil(s) defective. See Chapter 4.

11   Fuel/air mixture incorrect

1   Pilot screw(s) out of adjustment (Chapter 3).

2   Pilot jet or air passage clogged. Remove and overhaul the carburetors (Chapter 3).

3   Air bleed holes clogged. Remove carburetor and blow out all passages (Chapter 3).

4   Air cleaner clogged, poorly sealed or missing.

5   Air cleaner-to-carburetor boot poorly sealed. Look for cracks, holes or loose clamps and replace or repair defective parts.

6   Fuel level too high or too low. Adjust the floats (Chapter 3).

7   Fuel tank air vent obstructed. Make sure that the air vent passage in the filler cap is open.

8   Carburetor intake manifolds loose. Check for cracks, breaks, tears or loose clamps or bolts. Repair or replace the rubber boots.

12   Compression low

1   Spark plug loose. Remove the plug and inspect the threads. Reinstall and tighten to the specified torque (Chapter 1).

2   Cylinder head not sufficiently tightened down. If the cylinder head is suspected of being loose, then there's a chance that the gasket and head are damaged if the problem has persisted for any length of time. The head bolts should be tightened to the proper torque in the correct sequence (Chapter 2).

3   Improper valve clearance. This means that the valve is not closing completely and compression pressure is leaking past the valve. Check and adjust the valve clearances (Chapter 1).

4   Cylinder and/or piston worn. Excessive wear will cause compression pressure to leak past the rings. This is usually accompanied by worn rings as well. A top end overhaul is necessary (Chapter 2).

5   Piston rings worn, weak, broken, or sticking. Broken or sticking piston rings usually indicate a lubrication or carburetion problem that causes excess carbon deposits or seizures to form on the pistons and rings. Top end overhaul is necessary (Chapter 2).

6   Piston ring-to-groove clearance excessive. This is caused by excessive wear of the piston ring lands. Piston replacement is necessary (Chapter 2).

7   Cylinder head gasket damaged. If the head is allowed to become loose, or if excessive carbon build-up on the piston crown and combustion chamber causes extremely high compression, the head gasket may leak. Retorquing the head is not always sufficient to restore the seal, so gasket replacement is necessary (Chapter 2).

8   Cylinder head warped. This is caused by overheating or improperly tightened head bolts. Machine shop resurfacing or head replacement is necessary (Chapter 2).

9   Valve spring broken or weak. Caused by component failure or wear; the spring(s) must be replaced (Chapter 2).

10   Valve not seating properly. This is caused by a bent valve (from over-revving or improper valve adjustment), burned valve or seat (improper carburetion) or an accumulation of carbon deposits on the seat (from carburetion, lubrication problems). The valves must be cleaned and/or replaced and the seats serviced if possible (Chapter 2).

13   Poor acceleration

1   Carburetors leaking or dirty. Overhaul the carburetors (Chapter 3).

2   Timing not advancing. The pickup coil(s) or the igniter may be defective. If so, they must be replaced with new ones, as they can't be repaired.

3   Carburetors not synchronized. Adjust them with a vacuum gauge set or manometer (Chapter 1).
4   Engine oil viscosity too high. Using a heavier oil than that recommended in Chapter 1 can damage the oil pump or lubrication system and cause drag on the engine.
5   Brakes dragging. Usually caused by debris which has entered the brake piston sealing boot (disc brakes), weak springs (drum brakes), or from a warped disc or bent axle. Repair as necessary (Chapter 6).

Poor running or no power at high speed

14   Firing incorrect

1   Air filter restricted. Clean or replace filter (Chapter 1).
2   Spark plug fouled, defective or worn out. See Chapter 1 for spark plug maintenance.
3   Spark plug cap or secondary (HT) wiring defective. See Chapters 1 and 4 for details of the ignition system.
4   Spark plug cap not in good contact. See Chapter 4.
5   Incorrect spark plug. Wrong type, heat range or cap configuration. Check and install correct plugs listed in Chapter 1. A cold plug or one with a recessed firing electrode will not operate at low speeds without fouling.
6   Igniter defective. See Chapter 4.
7   Ignition coil(s) defective. See Chapter 4.

15   Fuel/air mixture incorrect

1   Main jet clogged. Dirt, water or other contaminants can clog the main jets. Clean the fuel tap filter, the float bowl area, and the jets and carburetor orifices (Chapter 3).
2   Main jet wrong size. The standard jetting is for sea level atmospheric pressure and oxygen content.
3   Throttle shaft-to-carburetor body clearance excessive. Refer to Chapter 3 for inspection and part replacement procedures.
4   Air bleed holes clogged. Remove and overhaul carburetors (Chapter 3).
5   Air cleaner clogged, poorly sealed, or missing.
6   Air cleaner-to-carburetor boot poorly sealed. Look for cracks, holes or loose clamps, and replace or repair defective parts.
7   Fuel level too high or too low. Adjust the float(s) (Chapter 3).
8   Fuel tank air vent obstructed. Make sure the air vent passage in the filler cap is open.
9   Carburetor intake manifolds loose. Check for cracks, breaks, tears or loose clamps or bolts. Repair or replace the rubber boots (Chapter 2).
10   Fuel tap clogged. Remove the tap and clean it and the filter (Chapter 1).
11   Fuel line clogged. Pull the fuel line loose and carefully blow through it.

16   Compression low

1   Spark plug loose. Remove the plug and inspect the threads. Reinstall and tighten to the specified torque (Chapter 1).
2   Cylinder head not sufficiently tightened down. If the cylinder head is suspected of being loose, then there's a chance that the gasket and head are damaged if the problem has persisted for any length of time. The head nuts should be tightened to the proper torque in the correct sequence (Chapter 2).
3   Improper valve clearance. This means that the valve is not closing completely and compression pressure is leaking past the valve. Check and adjust the valve clearances (Chapter 1).
4   Cylinder and/or piston worn. Excessive wear will cause compression

pressure to leak past the rings. This is usually accompanied by worn rings as well. A top end overhaul is necessary (Chapter 2).
5   Piston rings worn, weak, broken, or sticking. Broken or sticking piston rings usually indicate a lubrication or carburetion problem that causes excess carbon deposits or seizures to form on the pistons and rings. Top end overhaul is necessary (Chapter 2).
6   Piston ring-to-groove clearance excessive. This is caused by excessive wear of the piston ring lands. Piston replacement is necessary (Chapter 2).
7   Cylinder head gasket damaged. If the head is allowed to become loose, or if excessive carbon build-up on the piston crown and combustion chamber causes extremely high compression, the head gasket may leak. Retorquing the head is not always sufficient to restore the seal, so gasket replacement is necessary (Chapter 2).
8   Cylinder head warped. This is caused by overheating or improperly tightened head bolts. Machine shop resurfacing or head replacement is necessary (Chapter 2).
9   Valve spring broken or weak. Caused by component failure or wear; the spring(s) must be replaced (Chapter 2).
10   Valve not seating properly. This is caused by a bent valve (from over-revving or improper valve adjustment), burned valve or seat (improper carburetion) or an accumulation of carbon deposits on the seat (from carburetion or lubrication problems). The valves must be cleaned and/or replaced and the seats serviced if possible (Chapter 2).

17   Knocking or pinging

1   Carbon build-up in combustion chamber. Use of a fuel additive that will dissolve the adhesive bonding the carbon particles to the crown and chamber is the easiest way to remove the build-up. Otherwise, the cylinder head will have to be removed and decarbonized (Chapter 2).
2   Incorrect or poor quality fuel. Old or improper grades of fuel can cause detonation. This causes the piston to rattle, thus the knocking or pinging sound. Drain old fuel and always use the recommended fuel grade.
3   Spark plug heat range incorrect. Uncontrolled detonation indicates the plug heat range is too hot. The plug in effect becomes a glow plug, raising cylinder temperatures. Install the proper heat range plug (Chapter 1).
4   Improper air/fuel mixture. This will cause the cylinder to run hot, which leads to detonation. Clogged jets or an air leak can cause this imbalance. See Chapter 3.

18   Miscellaneous causes

1   Throttle valve doesn't open fully. Adjust the cable slack (Chapter 1).
2   Clutch slipping. May be caused by loose or worn clutch components. Refer to Chapter 2 for clutch overhaul procedures.
3   Timing not advancing.
4   Engine oil viscosity too high. Using a heavier oil than the one recommended in Chapter 1 can damage the oil pump or lubrication system and cause drag on the engine.
5   Brakes dragging. Usually caused by debris which has entered the brake piston sealing boot, or from a warped disc or bent axle. Repair as necessary.

Overheating

19   Engine overheats

1   Engine oil level low. Check and add oil (Chapter 1).

2   Wrong type of oil. If you're not sure what type of oil is in the engine, drain it and fill with the correct type (Chapter 1).
3   Air leak at carburetor intake boots. Check and tighten or replace as necessary (Chapter 3).
4   Float level low. Check and adjust if necessary (Chapter 3).
5   Worn oil pump or clogged oil passages. Check oil pressure (Chapter 2). Replace pump or clean passages as necessary.
6   Carbon build-up in combustion chambers. Use of a fuel additive that will dissolve the adhesive bonding the carbon particles to the piston crowns and chambers is the easiest way to remove the build-up. Otherwise, the cylinder head will have to be removed and decarbonized (Chapter 2).

20   Firing incorrect

1   Spark plugs fouled, defective or worn out. See Chapter 1 for spark plug maintenance.
2   Incorrect spark plugs.
3   Faulty ignition coil(s) (Chapter 4).

21   Fuel/air mixture incorrect

1   Main jet clogged. Dirt, water and other contaminants can clog the main jets. Clean the fuel tap filter, the float bowl area and the jets and carburetor orifices (Chapter 3).
2   Main jet wrong size. The standard jetting is for sea level atmospheric pressure and oxygen content.
3   Air cleaner poorly sealed or missing.
4   Air cleaner-to-carburetor boot poorly sealed. Look for cracks, holes or loose clamps and replace or repair.
5   Fuel level too low. Adjust the float(s) (Chapter 3).
6   Fuel tank air vent obstructed. Make sure that the air vent passage in the filler cap is open.
7   Carburetor intake manifolds loose. Check for cracks, breaks, tears or loose clamps or bolts. Repair or replace the rubber boots (Chapter 2).

22   Compression too high

1   Carbon build-up in combustion chamber. Use of a fuel additive that will dissolve the adhesive bonding the carbon particles to the piston crown and chamber is the easiest way to remove the build-up. Otherwise, the cylinder head will have to be removed and decarbonized (Chapter 2).
2   Improperly machined head surface or installation of incorrect gasket during engine assembly. Check Specifications (Chapter 2).

23   Engine load excessive

1   Clutch slipping. Can be caused by damaged, loose or worn clutch components. Refer to Chapter 2 for overhaul procedures.
2   Engine oil level too high. The addition of too much oil will cause pressurization of the crankcase and inefficient engine operation. Check Specifications and drain to proper level (Chapter 1).
3   Engine oil viscosity too high. Using a heavier oil than the one recommended in Chapter 1 can damage the oil pump or lubrication system as well as cause drag on the engine.
4   Brakes dragging. Usually caused by debris which has entered the brake piston sealing boot, or from a warped disc or bent axle. Repair as necessary.

24   Lubrication inadequate

1   Engine oil level too low. Friction caused by intermittent lack of lubrication or from oil that is overworked can cause overheating. The oil provides a definite cooling function in the engine. Check the oil level (Chapter 1).
2   Poor quality engine oil or incorrect viscosity or type. Oil is rated not only according to viscosity but also according to type. Some oils are not rated high enough for use in this engine. Check the Specifications section and change to the correct oil (Chapter 1).

25   Miscellaneous causes

Modification to exhaust system. Most aftermarket exhaust systems cause the engine to run leaner, which make them run hotter. When installing an accessory exhaust system, always reject the carburetors.

Clutch problems

26   Clutch slipping

1   Friction plates worn or warped. Overhaul the clutch assembly (Chapter 2).
2   Steel plates worn or warped (Chapter 2).
3   Clutch springs broken or weak. Old or heat-damaged (from slipping clutch) springs should be replaced with new ones (Chapter 2).
4   Worn or warped clutch plates. Replace (Chapter 2).
5   Clutch release mechanism defective. Replace any defective parts (Chapter 2).
6   Clutch hub or housing unevenly worn. This causes improper engagement of the discs. Replace the damaged or worn parts (Chapter 2).

27   Clutch not disengaging completely

1   Sticking cable. Inspect and lubricate or replace (Chapter 2).
2   Clutch plates warped or damaged. This will cause clutch drag, which in turn will cause the machine to creep. Overhaul the clutch assembly (Chapter 2).
3   Clutch spring tension uneven. Usually caused by a sagged or broken spring. Check and replace the spring (Chapter 2).
4   Engine oil deteriorated. Old, thin, worn out oil will not provide proper lubrication for the discs, causing the clutch to drag. Replace the oil and filter (Chapter 1).
5   Engine oil viscosity too high. Using a heavier oil than recommended in Chapter 1 can cause the plates to stick together, putting a drag on the engine. Change to the correct weight oil (Chapter 1).
6   Clutch housing seized on shaft. Lack of lubrication, severe wear or damage can cause the housing to seize on the shaft. Overhaul of the clutch, and perhaps transmission, may be necessary to repair the damage (Chapter 2).
7   Clutch release mechanism defective. Worn or damaged release mechanism parts can stick and fail to apply force to the pressure plate. Overhaul the release mechanism components (Chapter 2).
8   Loose clutch hub nut. Causes housing and hub misalignment putting a drag on the engine. Engagement adjustment continually varies. Overhaul the clutch assembly (Chapter 2).

Gear shifting problems

28 Doesn't go into gear or lever doesn't return

1 Clutch not disengaging. See Section 27.
2 Shift fork(s) bent or seized. Often caused by dropping the machine or from lack of lubrication. Overhaul the transmission (Chapter 2).
3 Gear(s) stuck on shaft. Most often caused by a lack of lubrication or excessive wear in transmission bearings and bushings. Overhaul the transmission (Chapter 2).
4 Shift cam binding. Caused by lubrication failure or excessive wear. Replace the cam and bearings (Chapter 2).
5 Shift lever return spring weak or broken (Chapter 2).
6 Shift lever broken. Splines stripped out of lever or shaft, caused by allowing the lever to get loose or from dropping the machine. Replace necessary parts (Chapter 2).
7 Shift mechanism pawl broken or worn. Full engagement and rotary movement of shift cam results. Replace shaft assembly (Chapter 2).
8 Pawl spring broken. Allows pawl to float, causing sporadic shift operation. Replace spring (Chapter 2).

29 Jumps out of gear

1 Shift fork(s) worn. Overhaul the transmission (Chapter 2).
2 Gear groove(s) worn. Overhaul the transmission (Chapter 2).
3 Gear dogs or dog slots worn or damaged. The gears should be inspected and replaced. No attempt should be made to service the worn parts.

30 Overshifts

1 Pawl spring weak or broken (Chapter 2).
2 Shift cam stopper lever not functioning (Chapter 2).
3 Overshift limiter broken or distorted (Chapter 2).

Abnormal engine noise

31 Knocking or pinging

1 Carbon build-up in combustion chamber. Use of a fuel additive that will dissolve the adhesive bonding the carbon particles to the piston crown and chamber is the easiest way to remove the build-up. Otherwise, the cylinder head will have to be removed and decarbonized (Chapter 2).
2 Incorrect or poor quality fuel. Old or improper fuel can cause detonation. This causes the pistons to rattle, thus the knocking or pinging sound. Drain the old fuel and always use the recommended grade fuel (Chapter 3).
3 Spark plug heat range incorrect. Uncontrolled detonation indicates that the plug heat range is too hot. The plug in effect becomes a glow plug, raising cylinder temperatures. Install the proper heat range plug (Chapter 1).
4 Improper air/fuel mixture. This will cause the cylinders to run hot and lead to detonation. Clogged jets or an air leak can cause this imbalance. See Chapter 3.

32 Piston slap or rattling

1 Cylinder-to-piston clearance excessive. Caused by improper assembly. Inspect and overhaul top end parts (Chapter 2).
2 Connecting rod bent. Caused by over-revving, trying to start a badly flooded engine or from ingesting a foreign object into the combustion chamber. Replace the damaged parts (Chapter 2).
3 Piston pin or piston pin bore worn or seized from wear or lack of lubrication. Replace damaged parts (Chapter 2).
4 Piston ring(s) worn, broken or sticking. Overhaul the top end (Chapter 2).
5 Piston seizure damage. Usually from lack of lubrication or overheating. Replace the pistons and bore the cylinders, as necessary (Chapter 2).
6 Connecting rod upper or lower end clearance excessive. Caused by excessive wear or lack of lubrication. Replace worn parts.

33 Valve noise

1 Incorrect valve clearances. Adjust the clearances by referring to Chapter 1.
2 Valve spring broken or weak. Check and replace weak valve springs (Chapter 2).
3 Camshaft or cylinder head worn or damaged. Lack of lubrication at high rpm is usually the cause of damage. Insufficient oil or failure to change the oil at the recommended intervals are the chief causes. Since there are no replaceable bearings in the head, the head itself will have to be replaced if there is excessive wear or damage (Chapter 2).

34 Other noise

1 Cylinder head gasket leaking.
2 Exhaust pipe leaking at cylinder head connection. Caused by improper fit of pipe(s) or loose exhaust flange. All exhaust fasteners should be tightened evenly and carefully. Failure to do this will lead to a leak.
3 Crankshaft runout excessive. Caused by a bent crankshaft (from over-revving) or damage from an upper cylinder component failure. Can also be attributed to dropping the machine on either of the crankshaft ends.
4 Engine mounting bolts loose. Tighten all engine mount bolts to the specified torque (Chapter 2).
5 Crankshaft bearings worn (Chapter 2).
6 Camshaft chain tensioner out of adjustment or defective. Adjust (Chapter 1) or replace (Chapter 2).
7 Camshaft chain, sprockets or guides worn (Chapter 2).

Abnormal driveline noise

35 Clutch noise

1 Clutch housing/friction plate clearance excessive (Chapter 2).
2 Loose or damaged clutch pressure plate and/or bolts (Chapter 2).

36 Transmission noise

1 Bearings worn. Also includes the possibility that the shafts are worn. Overhaul the transmission (Chapter 2).
2 Gears worn or chipped (Chapter 2).
3 Metal chips jammed in gear teeth. Probably pieces from a broken clutch, gear or shift mechanism that were picked up by the gears. This will cause early bearing failure (Chapter 2).
4 Engine oil level too low. Causes a howl from transmission. Also affects engine power and clutch operation (Chapter 1).

37 Final drive noise

1   Chain not adjusted properly (Chapter 1).
2   Engine sprocket or rear sprocket loose. Tighten fasteners (Chapters 2 and 5).
3   Sprocket(s) worn. Replace sprocket(s). (Chapter 5).
4   Rear sprocket warped. Replace (Chapter 5).
5   Wheel coupling worn. Replace coupling (Chapter 5).

Abnormal frame and suspension noise

38 Front end noise

1   Low fluid level or improper viscosity oil in forks. This can sound like spurting and is usually accompanied by irregular fork action (Chapter 5).
2   Spring weak or broken. Makes a clicking or scraping sound. Fork oil, when drained, will have a lot of metal particles in it (Chapter 5).
3   Steering head bearings loose or damaged. Clicks when braking. Check and adjust or replace as necessary (Chapter 5).
4   Fork clamps loose. Make sure all fork clamp pinch bolts are tight (Chapter 5).
5   Fork tube bent. Good possibility if machine has been dropped. Replace tube with a new one (Chapter 5).
6   Front axle or axle clamp bolt loose. Tighten them to the specified torque (Chapter 6).

39 Shock absorber noise

1   Fluid level incorrect. Indicates a leak caused by defective seal. Shock will be covered with oil. Replace shock (Chapter 5).
2   Defective shock absorber with internal damage. This is in the body of the shock and Can't be remedied. The shock must be replaced with a new one (Chapter 5).
3   Bent or damaged shock body. Replace the shock with a new one (Chapter 5).

40 Brake noise

1   Squeal caused by pad shim not installed or positioned correctly (Chapter 6).
2   Squeal caused by dust on brake pads or shoes. Usually found in combination with glazed pads. Clean using brake cleaning solvent (Chapter 6).
3   Contamination of brake pads or shoes. Oil, brake fluid or dirt causing brake to chatter or squeal. Clean or replace pads or shoes (Chapter 6).
4   Pads glazed. Caused by excessive heat from prolonged use or from contamination. Do not use sandpaper, emery cloth, carborundum cloth or any other abrasive to roughen the pad surfaces as abrasives will stay in the pad material and damage the disc. A very fine flat file can be used, but pad replacement is suggested as a cure (Chapter 6).
5   Disc warped. Can cause a chattering, clicking or intermittent squeal. Usually accompanied by a pulsating lever and uneven braking. Replace the disc (Chapter 6).
6   Drum out-of-round. Can cause chattering, clicking or intermittent squeal. Usually accompanied by a pulsating pedal and uneven braking. Refinish or replace drum (Chapter 6).
7   Loose or worn wheel bearings. Check and replace as needed (Chapter 6).

Oil level indicator light comes on

41 Engine lubrication system

1   These models use an oil level light rather than an oil pressure light.
2   Engine oil level low. Inspect for leak or other problem causing low oil level and add recommended oil (Chapters 1 and 2).

42 Electrical system

1   Oil level switch defective. Check the switch according to the procedure in Chapter 8. Replace it if it's defective.
2   Oil level indicator light circuit defective. Check for pinched, shorted, disconnected or damaged wiring (Chapter 8).

Excessive exhaust smoke

43 White smoke

1   Piston oil ring worn. The ring may be broken or damaged, causing oil from the crankcase to be pulled past the piston into the combustion chamber. Replace the rings with new ones (Chapter 2).
2   Cylinders worn, cracked, or scored. Caused by overheating or oil starvation. The cylinders will have to be rebored and new pistons installed.
3   Valve oil seal damaged or worn. Replace oil seals with new ones (Chapter 2).
4   Valve guide worn. Perform a complete valve job (Chapter 2).
5   Engine oil level too high, which causes the oil to be forced past the rings. Drain oil to the proper level (Chapter 1).
6   Head gasket broken between oil return and cylinder. Causes oil to be pulled into the combustion chamber. Replace the head gasket and check the head for warpage (Chapter 2).
7   Abnormal crankcase pressurization, which forces oil past the rings. Clogged breather or hoses usually the cause (Chapter 3).

44 Black smoke

1   Air cleaner clogged. Clean or replace the element (Chapter 1).
2   Main jet too large or loose. Compare the jet size to the Specifications (Chapter 3).
3   Choke stuck, causing fuel to be pulled through choke circuit (Chapter 3).
4   Fuel level too high. Check and adjust the float level as necessary (Chapter 3).
5   Inlet needle held off needle seat. Clean the float bowls and fuel line and replace the needles and seats if necessary (Chapter 3).

45 Brown smoke

1   Main jet too small or clogged. Lean condition caused by wrong size main jet or by a restricted orifice. Clean float bowl and jets and compare jet size to Specifications (Chapter 3).
2   Fuel flow insufficient. Fuel inlet needle valve stuck closed due to chemical reaction with old fuel. Float level incorrect. Restricted fuel line. Clean line and float bowl and adjust floats if necessary.
3   Carburetor intake manifolds loose (Chapter 3).
4   Air cleaner poorly sealed or not installed (Chapter 1).

Poor handling or stability

46 Handlebar hard to turn

1 Steering stem locknut too tight (Chapter 5).
2 Bearings damaged. Roughness can be felt as the bars are turned from side-to-side. Replace bearings and races (Chapter 5).
3 Races dented or worn. Denting results from wear in only one position (e.g., straight ahead), from a collision or hitting a pothole or from dropping the machine. Replace races and bearings (Chapter 5).
4 Steering stem lubrication inadequate. Causes are grease getting hard from age or being washed out by high pressure car washes. Disassemble steering head and repack bearings (Chapter 5).
5 Steering stem bent. Caused by a collision, hitting a pothole or by dropping the machine. Replace damaged part. Don't try to straighten the steering stem (Chapter 5).
6 Front tire air pressure too low (Chapter 1).

47 Handlebar shakes or vibrates excessively

1 Tires worn or out of balance (Chapter 6).
2 Swingarm bearings worn. Replace worn bearings by referring to Chapter 6.
3 Rim(s) warped or damaged. Inspect wheels for runout (Chapter 6).
4 Wheel bearings worn. Worn front or rear wheel bearings can cause poor tracking. Worn front bearings will cause wobble (Chapter 6).
5 Handlebar clamp bolts loose (Chapter 5).
6 Steering stem or fork clamps loose. Tighten them to the specified torque (Chapter 5).
7 Engine mounting bolts loose. Will cause excessive vibration with increased engine rpm (Chapter 2).

48 Handlebar pulls to one side

1 Frame bent. Definitely suspect this if the machine has been dropped. May or may not be accompanied by cracking near the bend. Replace the frame (Chapter 5).
2 Wheel out of alignment. Caused by improper location of axle spacers or from bent steering stem or frame (Chapter 5).
3 Swingarm bent or twisted. Caused by age (metal fatigue) or impact damage. Replace the arm (Chapter 5).
4 Steering stem bent. Caused by impact damage or by dropping the motorcycle. Replace the steering stem (Chapter 5).
5 Fork leg bent. Disassemble the forks and replace the damaged parts (Chapter 6).
6 Fork oil level uneven. Check and add or drain as necessary (Chapter 1).

49 Poor shock absorbing qualities

1 Too hard:
 a) Fork oil level excessive (Chapter 5).
 b) Fork oil viscosity too high. Use a lighter oil (see the Specifications in Chapter 1).
 c) Fork tube bent. Causes a harsh, sticking feeling (Chapter 5).
 d) Shock shaft or body bent or damaged (Chapter 5).
 e) Fork internal damage (Chapter 5).
 f) Shock internal damage.
 g) Tire pressure too high (Chapters 1 and 6).
2 Too soft:
 a) Fork or shock oil insufficient and/or leaking (Chapter 5).

 b) Fork oil level too low (Chapter 5).
 c) Fork oil viscosity too light (Chapter 5).
 d) Fork springs weak or broken (Chapter 5).

Braking problems

50 Brakes are spongy, don't hold

1 Air in brake line. Caused by inattention to master cylinder fluid level or by leakage. Locate problem and bleed brakes (Chapter 6).
2 Pad or disc worn (Chapters 1 and 6).
3 Brake fluid leak. See paragraph 1.
4 Contaminated pads. Caused by contamination with oil, grease, brake fluid, etc. Clean or replace pads. Clean disc thoroughly with brake cleaner (Chapter 6).
5 Brake fluid deteriorated. Fluid is old or contaminated. Drain system, replenish with new fluid and bleed the system (Chapter 6).
6 Master cylinder internal parts worn or damaged causing fluid to bypass (Chapter 6).
7 Master cylinder bore scratched by foreign material or broken spring. Repair or replace master cylinder (Chapter 6).
8 Disc warped. Replace disc (Chapter 6).

51 Brake lever or pedal pulsates

1 Disc warped. Replace disc (Chapter 6).
2 Axle bent. Replace axle (Chapter 5).
3 Brake caliper bolts loose (Chapter 6).
4 Brake caliper shafts damaged or sticking, causing caliper to bind. Lube the shafts or replace them if they are corroded or bent (Chapter 6).
5 Wheel warped or otherwise damaged (Chapter 6).
6 Wheel bearings damaged or worn (Chapter 6).
7 Drum out-of-round (Radian models). Refinish or replace drum (Chapter 6).

52 Brakes drag

1 Master cylinder piston seized. Caused by wear or damage to piston or cylinder bore (Chapter 6).
2 Lever balky or stuck. Check pivot and lubricate (Chapter 6).
3 Brake caliper binds. Caused by inadequate lubrication or damage to caliper shafts (Chapter 6).
4 Brake caliper piston seized in bore. Caused by wear or ingestion of dirt past deteriorated seal (Chapter 6).
5 Brake pad damaged. Pad material separated from backing plate. Usually caused by faulty manufacturing process or from contact with chemicals. Replace pads (Chapter 6).
6 Pads or shoes improperly installed (Chapter 6).
7 Rear brake pedal free play insufficient (Chapter 1).
8 Rear brake springs weak Chapter 6).

Electrical problems

53 Battery dead or weak

1 Battery faulty. Caused by sulfated plates which are shorted through sedimentation or low electrolyte level. Also, broken battery terminal making only occasional contact (Chapter 8).
2 Battery cables making poor contact (Chapter 8).

3   Load excessive. Caused by addition of high wattage lights or other electrical accessories.

4   Ignition switch defective. Switch either grounds internally or fails to shut off system. Replace the switch (Chapter 8).

5   Regulator/rectifier defective (Chapter 8).

6   Stator coil open or shorted (Chapter 8).

7   Wiring faulty. Wiring grounded or connections loose in ignition, charging or lighting circuits (Chapter 8).

54   Battery overcharged

1   Regulator/rectifier defective. Overcharging is noticed when battery gets excessively warm or boils over (Chapter 8).

2   Battery defective. Replace battery with a new one (Chapter 8).

3   Battery amperage too low, wrong type or size. Install manufacturer's specified amp-hour battery to handle charging load (Chapter 8).

Chapter 1
Tune-up and routine maintenance

Contents

Specifications

Engine

Spark plugs
  Type
    US models .... NGK D8EA or ND X24ES-U
    UK models .... NGK DR8ES-L or ND X24ESR-U
  Gap   .... 0.6 to 0.7 mm (0.024 to 0.028 inch)
Valve clearances (COLD engine)
  Intake.. 0.11 to 0.15 mm (0.004 to 0.005 inch)
  Exhaust .... 0.16 to 0.20 mm (0.006 to 0.008 inch)
Engine idle speed
  YX600 Radian... 1300 +/- 50 rpm
  All others .... 1200 +/- 50 rpm
Cylinder compression pressure (at sea level)
  Standard... 10.6 Bars (156 psi)
  Maximum.. 11.2 Bars (164 psi)
  Minimum... 9.7 Bars (142 psi)
Carburetor synchronization
  Vacuum at idle speed... 175 +/- 5 mm Hg (6.89 +/- 0.2 inch Hg)
  Maximum vacuum difference between cylinders............. 10 mm Hg (0.39 inch Hg)
Cylinder numbering (from left side to right side of bike)......... 1-2-3-4

Miscellaneous

Disc brake pad minimum thickness (front and rear)............. 0.5 mm (.020 inch)
Drum brake shoe lining minimum thickness......................... 2 mm (.080 inch)
Brake pedal position (below top of footpeg)
  FJ600 .... 30 mm (1.20 inches)
  FZ600 .... 40 mm (1.6 inches)
  XJ600 .... 30 mm (1.2 inches)
  YX600 Radian.. 15 mm (0.6 inch)
Freeplay adjustments
  Throttle grip
    FJ600 .... 3 to 7 mm (0.12 to 0.28 inch)
    FZ600 .... 2 to 5 mm (0.08 to 0.20 inch
    XJ600 .... 3 to 7 mm (0.12 to 0.28 inch)
    YX600 Radian .... 2 to 5 mm (0.08 to 0.20 inch)
  Clutch lever
    FJ600... 2 to 3 mm (0.8 to 1.2 inch)
    FZ600... 8 to 12 mm (0.3 to 0.5 inch)
    XJ600... 2 to 3 mm (0.8 to 1.2 inch)
    YX600 Radian .... 10 to 15 mm (0.4 to 0.6 inch)
  Front brake lever
    FJ600... 5 to 8 mm (0.2 to 0.3 inch)
    FZ600... 0 to 1 mm (0 to 0.04 inch)
    XJ600... 5 to 8 mm (0.2 to 0.3 inch)
    YX600 Radian .... 2 to 5 mm (0.08 to 0.20 inch)
  Brake pedal
    FJ600... 20 to 30 mm (0.80 to 1.20 inch)
    FZ600... 13 to 15 mm (0.51 to 0.59 inch)
    XJ600... 20 to 30 mm (0.80 to 0.12 inch)
    YX600 Radian .... 20 to 30 mm (0.80 to 1.20 inch)
Drive chain slack.. 20 to 30 mm (0.8 to 1.2 inch)
Battery electrolyte specific gravity .... 1.280 at 20-degrees C (68-degrees F)
Minimum tire tread depth*... 1 mm (0.040 inch)
Tire pressures (cold)
  FJ600, XJ600
    Front
      Up to 90 kg (198 lbs) .... 1.77 Bars (26 psi)
      Above 90 kg (198 lbs) or high speed riding............... 1.9 Bars (28 psi)
    Rear
      Up to 90 kg (198 lbs) .... 1.9 Bars (28 psi)
      Above 90 kg (198 lbs) or high speed riding............... 2.2 Bars (32 psi)
  FZ600
    Front
      Up to 90 kg (198 lbs) .... 1.77 Bars (26 psi)
      Above 90 kg (198 lbs) or high speed riding............... 1.9 Bars (28 psi)

Rear
    Up to 90 kg (198 lbs) .... 1.9 Bars (28 psi)
    Above 90 kg (198 lbs) .... 2.48 Bars (36 psi)
    High speed riding .... 2.2 Bars (32 psi)
YX600 Radian
  Front
    Up to 90 kg (198 lbs) .... 1.77 Bars (26 psi)
    Above 90 kg (198 lbs) or high speed riding .... 1.9 Bars (28 psi)
  Rear
    Up to 90 kg (198 lbs) .... 1.9 Bars (28 psi)
    90 to 160 kg (198 to 353 lbs) or high speed riding................. 2.2 Bars (32 psi)
    Above 160 kg (353 lbs) .... 2.48 Bars (36 psi)

Recommended lubricants and fluids

Engine/transmission oil

Type .... API grade SE, SF or SG
Viscosity
  Up to 15 degrees C (60 degrees F) .... SAE 10W30
  Above 5 degrees C (40 degrees F)... SAE 20W40
Capacity
  With filter change
    FJ600, FZ600, XJ600 .... 2.6 liters (2.7 US qts, 4.6 Imperial pts)
    YX600 Radian .... 2.5 liters (2.6 US qts, 4.4 Imperial pts)
  Oil change only.. 2.2 liters (2.3 US qts, 3.8 Imperial pts)
Brake fluid .... DOT 4 (DOT 3 may be used if DOT 4 is unavailable)

Fork oil

Type .... SAE 10W - fork oil
Capacity
  FJ600 .... 287 cc (9.7 US fl oz, 10.1 Imperial fl oz)
  FZ600 .... 315 cc (10.7 US fl oz, 11.1 Imperial fl oz)
  XJ600 .... 269 cc (9.09 US fl oz, 9.47 Imperial fl oz)
  YX600 Radian.. 320 cc (10.8 US oz, 11.3 Imperial oz)
Oil level**
  FJ600 .... Not specified
  FZ600 .... 117 mm (4.61 inches)
  XJ600
    1984 and 1985 .... Not specified
    1989-on.. 149 cc (5.87 inches)
  YX600 Radian.. not specified

Miscellaneous

Drive chain... SAE 30 to 50W engine oil
Wheel bearings .... Medium weight, lithium-based multi-purpose grease
Swingarm pivot bearings .... Medium weight, lithium-based waterproof wheel bearing grease
Cables and lever pivots .... Chain and cable lubricant or 10W30 motor oil
Sidestand/centerstand pivots .... Medium-weight, lithium-based multi-purpose grease
Brake pedal/shift lever pivots .... Chain and cable lubricant or 10W30 motor oil
Throttle grip .... Multi-purpose grease or dry film lubricant

Torque specifications

Oil drain plug .... 43 Nm (31 ft-lbs)
Oil filter bolt .... 15 Nm (11 ft-lbs)
Oil filter drain screw .... 7 Nm (5.1 ft-lbs)
Spark plugs.. 17.5 Nm (12.5 ft-lbs)
Steering head bearing ring nut
  Initial torque... 38 Nm (27 ft-lbs)***
  Final torque .... See text
Valve cover bolts .... See Chapter 2

*In the UK, tread depth must be at least 1 mm over 3/4 of the tread breadth all the way around the tire, with no bald patches.
**With the forks fully compressed and the spring removed.
***Using Yamaha ring nut wrench YU-33975 and a torque wrench placed at a right angle to the ring nut wrench.

1 Yamaha FJ, FZ, XJ, YX Radian 600 Routine maintenance intervals

Note: *The pre-ride inspection outlined in the owner's manual covers checks and maintenance that should be carried out on a daily basis. it's condensed and included here to remind you of its importance. Always perform the pre-ride inspection at every maintenance interval (in addition to the procedures listed). The intervals listed below are the shortest intervals recommended by the manufacturer for each particular operation during the model years covered in this manual. Your owner's manual may have different intervals for your model.*

Daily or before riding

Check the engine oil level
Check the fuel level and inspect for leaks
Check the operation of both brakes - also check the fluid level and look for leakage
Check the tires for damage, the presence of foreign objects and correct air pressure
Check the throttle for smooth operation and correct freeplay
Check the operation of the clutch - make sure the freeplay is correct
Make sure the steering operates smoothly, without looseness and without binding
Check for proper operation of the headlight, taillight, brake light, turn signals, indicator lights, speedometer and horn
Make sure the sidestand and centerstand (if equipped) return to their fully up positions and stay there under spring pressure
Make sure the engine kill switch works properly

After the initial 500 miles

Perform all of the daily checks plus:
Check/adjust the idle speed and carburetor synchronization
Check/adjust the drive chain slack
Change the engine oil and oil filter
Check the tightness of all fasteners
Check/adjust the clutch lever position
Check the brake fluid level
Check/adjust the brake pedal position
Check the operation of the brake light
Check the operation of the sidestand switch
Adjust cam chain tension
Lubricate the throttle, clutch, speedometer and choke cables (if equipped)

Every 500 miles

Check/adjust the drive chain slack
Lubricate the drive chain

Every 6000 km/4000 miles or 6 months

Change the engine oil
Adjust the cam chain tension

Clean the air filter element and replace it if necessary
Clean and gap the spark plugs
Check/adjust the idle speed
Check/adjust the carburetor synchronization
Check the brake fluid level
Lubricate brake pad edges and caliper cavities*
Check the drum brake operation
Check the brake discs and pads
Check/adjust the brake pedal position
Check the operation of the brake light
Lubricate the clutch and brake lever pivots
Lubricate the shift/brake lever pivots and the sidestand/centerstand pivots
Check the steering
Check the front forks for proper operation and fluid leaks
Inspect and lubricate the rear suspension adjusting belt (if equipped)
Check the tires and wheels
Check the battery electrolyte level
Check the exhaust system for leaks and check the tightness of the fasteners
Check the cleanliness of the fuel system and the condition of the fuel lines and vacuum hoses
Inspect the crankcase ventilation system
Check the operation of the sidestand switch
Lubricate the throttle cables
Required in the UK and recommended wherever salt is used on the roads

Every 12,000 km/8,000 miles or 12 months

All of the items above plus:
Change the engine oil and oil filter
Replace the spark plugs
Replace the alternator brushes (if equipped)
Check/adjust the valve clearances

Every 18,000 km/12,000 miles or 18 months

All of the items above plus:
Inspect the evaporative emission control system (if equipped)

Every 24,000 km/16,000 miles or two years

Change the brake fluid
Clean and repack the steering head bearings
Lubricate the swingarm bearings and rear suspension pivot points

2.3a Maintenance information printed on decals includes tune-up and lubrication data . . .

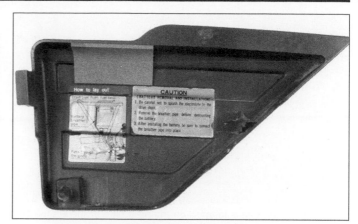

2.3b . . . battery vent tube location . . .

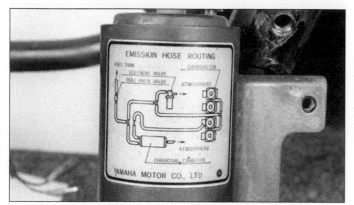

2.3c . . . and vacuum hose routing information

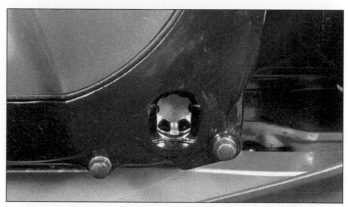

3.3 Check oil level at the inspection window; it should be between the Minimum and Maximum marks

2 Introduction to tune-up and routine maintenance

Refer to illustrations 2.3a, 2.3b and 2.3c

This Chapter covers in detail the checks and procedures necessary for the tune-up and routine maintenance of your motorcycle. Section 1 includes the routine maintenance schedule, which is designed to keep the machine in proper running condition and prevent possible problems. The remaining Sections contain detailed procedures for carrying out the items listed on the maintenance schedule, as well as additional maintenance information designed to increase reliability.

Since routine maintenance plays such an important role in the safe and efficient operation of your motorcycle, it is presented here as a comprehensive check list. For the rider who does all his/her own maintenance, these lists outline the procedures and checks that should be done on a routine basis.

Maintenance information is printed on decals attached to the motorcycle **(see illustrations)**. If the information on the decals differs from that included here, use the information on the decal.

Deciding where to start or plug into the routine maintenance schedule depends on several factors. If you have a motorcycle whose warranty has recently expired, and if it has been maintained according to the warranty standards, you may want to pick up routine maintenance as it coincides with the next mileage or calendar interval. If you have owned the machine for some time but have never performed any maintenance on it, then you may want to start at the nearest interval and include some additional procedures to ensure that nothing important is overlooked. If you have just had a major engine overhaul, then you may want to start the maintenance routine from the beginning. If you have a used machine and have no knowledge of its history or maintenance record, you may desire to combine all the checks into one large service initially and then settle into the maintenance schedule prescribed.

The Sections which outline the inspection and maintenance procedures are written as step-by-step comprehensive guides to the performance of the work. They explain in detail each of the routine inspections and maintenance procedures on the check list. References to additional information in applicable Chapters is also included and should not be overlooked.

Before beginning any maintenance or repair, the machine should be cleaned thoroughly, especially around the oil filter, spark plugs, cylinder head covers, side covers, carburetors, etc. Cleaning will help ensure that dirt does not contaminate the engine and will allow you to detect wear and damage that could otherwise easily go unnoticed.

3 Fluid levels - check

Engine oil

Refer to illustrations 3.3 and 3.4

1 Start the engine and allow it to reach normal operating temperature. **Warning:** *Do not run the engine in an enclosed space such as a garage or shop.*

2 Stop the engine and allow the machine to sit undisturbed in a level position for about five minutes. Place the bike on the centerstand (if equipped) or prop it securely in an upright position.

3 With the engine off, check the oil level in the window located at the lower part of the right crankcase cover. The oil level should be between the Maximum and Minimum level marks next to the window **(see illustration)**.

3.4 The oil filler cap is located on the right side of the engine on the clutch cover (arrow)

3.7a With the front master cylinder in a level position, check fluid level in the inspection window (arrow)

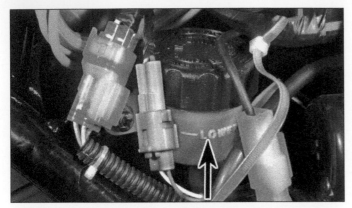

3.7b The fluid level in the rear brake master cylinder can be checked by looking through the plastic reservoir - fluid must be above the lower mark (arrow)

3.9 To add fluid to the front brake reservoir, remove the master cylinder cover screws (arrows) and lift off the cover and diaphragm

4   If the level is below the Minimum mark, remove the oil filler cap from the right side of the crankcase **(see illustration)** and add enough oil of the recommended grade and type to bring the level up to the Maximum mark. Do not overfill.

Brake fluid

Refer to illustrations 3.7a, 3.7b and 3.9

5   In order to ensure proper operation of the hydraulic disc brakes, the fluid level in the master cylinder reservoir must be properly maintained.

6   With the motorcycle on the centerstand or held upright, turn the handlebars until the top of the front master cylinder is as level as possible. If necessary, tilt the motorcycle make it level.

7   Look closely at the inspection window in the master cylinder reservoir (front brakes) or at the level marks on the reservoir (rear brakes). Make sure that the fluid level is above the Lower mark on the reservoir **(see illustrations).**

8   If the level is low, the fluid must be replenished. Before removing the master cylinder cover or cap, cover the fuel tank to protect it from brake fluid spills (which will damage the paint) and remove all dust and dirt from the area around the cover or cap.

9   To top up front brake fluid, remove the screws **(see illustration)** and lift off the cover and rubber diaphragm. To top up rear brake fluid, unscrew the cap from the reservoir. **Note:** *Do not operate the brakes with the cover or cap removed.*

10   Add new, clean brake fluid of the recommended type until the level is above the inspection window in the front reservoir or the lower line in the rear reservoir. Do not mix different brands of brake fluid in the reservoir, as they may not be compatible.

11   On front brakes, install the rubber diaphragm and the cover. Tighten the screws evenly, but do not overtighten them. On rear brakes, install the reservoir cap.

12   Wipe any spilled fluid off the reservoir body and reposition and tighten the brake lever and master cylinder assembly if it was moved.

13   If the brake fluid level was low, inspect the brake system for leaks.

4   Battery electrolyte level/specific gravity - check

Refer to illustrations 4.4, 4.7, 4.11a, 4.11b and 4.11c

Warning: *Be extremely careful when handling or working around the battery. The electrolyte is very caustic and an explosive gas (hydrogen) is given off when the battery is charging.* **Note:** *The first Steps describe battery removal. If the electrolyte level is known to be sufficient it won't be necessary to remove the battery.*

1   This procedure applies to original equipment batteries that have removable filler caps, which can be removed to add water to the battery. Sealed maintenance-free batteries can't be topped up.

2   Remove the seat (see Chapter 7).

3   If you're working on a Radian, remove the right side cover (see Chapter 7).

4   Remove the screws securing the battery cables to the battery terminals (remove the negative cable first, positive cable last) **(see illustration).** Remove the battery securing strap and pull the battery straight up to remove it. The electrolyte level will now be visible through the translucent battery case - it should be between the Upper and Lower level marks.

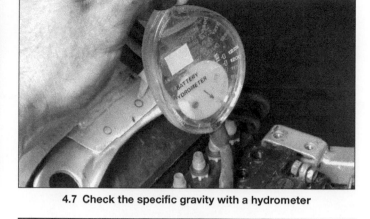

4.4 Detach the negative cable (right arrow) from the battery first, then detach the positive cable (left arrow); the plastic cap protects the positive terminal from accidental contact with metal

4.7 Check the specific gravity with a hydrometer

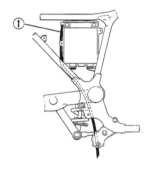

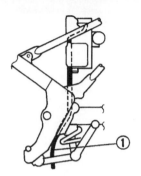

4.11a Be sure the battery vent tube is routed correctly, with no kinking or obstructions; this is an FJ/XJ600 . . .

4.11b . . . this is an FZ600 . . .

5   If the electrolyte is low, remove the cell caps and fill each cell to the upper level mark with distilled water only. Do not use tap water (except in an emergency), and do not overfill. The cell holes are quite small, so it may help to use a plastic squeeze bottle with a small spout to add the water. If the level is within the marks on the case, additional water is not necessary.

6   Next, check the specific gravity of the electrolyte in each cell with a small hydrometer made especially for motorcycle batteries. These are available from most dealer parts departments or motorcycle accessory stores.

7   Remove the caps, draw some electrolyte from the first cell into the hydrometer **(see illustration)** and note the specific gravity. Compare the reading to the Specifications listed in this Chapter. **Note:** *Add 0.004 points to the reading for every 10-degrees F above 20-degrees C (68-degrees F) - subtract 0.004 points from the reading for every 10-degrees below 20-degrees C (68-degrees F). Return the electrolyte to the appropriate cell and repeat the check for the remaining cells. When the check is complete, rinse the hydrometer thoroughly with clean water.*

8   If the specific gravity of the electrolyte in each cell is as specified, the battery is in good condition and is apparently being charged by the machine's charging system.

9   If the specific gravity is low, the battery is not fully charged. This may be due to corroded battery terminals, a dirty battery case, a malfunctioning charging system, or loose or corroded wiring connections. On the other hand, it may be that the battery is worn out, especially if the machine is old, or that infrequent use of the motorcycle prevents normal charging from taking place.

10   Be sure to correct any problems and charge the battery if necessary. Refer to Chapter 8 for additional battery maintenance and charging procedures.

11   Install the battery cell caps, tightening them securely. Reconnect the cables to the battery, attaching the positive cable first and the negative

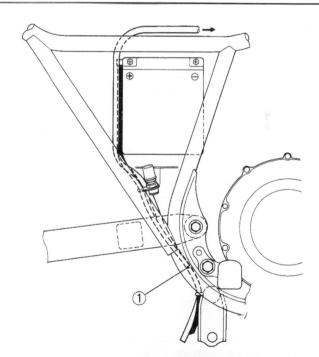

4.11c . . . and this is a YX600 Radian

cable last. Make sure to install the insulating boot over the positive terminal. Install the fuel tank mount and the seat. Be very careful not to pinch or otherwise restrict the battery vent tube, as the battery may build up enough internal pressure during normal charging system operation to explode **(see illustrations)**. Additionally, battery acid spilled onto the frame can cause accelerated corrosion and metal fatigue.

5.2a Lift off the front pad cover to expose the pads . . .

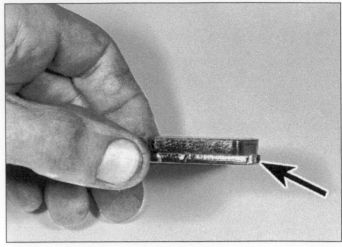

5.2b . . . and check to see if the raised corner of the pad backing metal (arrow) is close to the disc; if it is, the pad is worn and the full set of pads must be replaced (pad removed for clarity)

5   Brake pads - check

Refer to illustrations 5.2a, 5.2b and 5.4

1   The front and rear brake pads should be checked at the recommended intervals and replaced with new ones when worn beyond the limit described in this Section.

2   To check the front brake pads, remove the pad cover **(see illustration)**. Reach up and operate the brake lever while you look at the back of the caliper. If the pad wear indicator is close to the disc **(see illustration)**, the pads are worn excessively and must be replaced with new ones (see Chapter 6).

3   Repeat Step 2 to check the other front caliper.

4   To check the rear brake pads, have an assistant press the brake pedal firmly while you look at the pads through the back of the caliper **(see illustration)**. Press the brake pedal firmly while you look at the wear indicators on the pads **(see illustration 5.2b)**. If the indicators are close to the disc, replace the pads (see Chapter 6).

5   If the pads are in good condition, reinstall the covers (front brakes only). The words "Uncover for pad service" stamped in the covers may be upside down when the cover is installed. This doesn't mean the cover is upside down.

6   On UK models, remove the brake pads and lubricate the edges of the pad backing plates and the pad cavities in the calipers with special lubricants (see *Brake pads - replacement* in Chapter 6).

6   Brake shoes - check

Refer to illustration 6.1

1   Check rear brake shoe wear by pressing the brake pedal and observing the movement of the wear indicator pointer **(see illustration)**.

2   If the arrow moves out of the safe range, replace the rear brake shoes (see Chapter 6).

7   Brake system - general check

Refer to illustration 7.7

1   A routine general check of the brakes will ensure that any problems are discovered and remedied before the rider's safety is jeopardized.

2   Check the brake lever and pedal for loose connections, excessive play, bends, and other damage. Replace any damaged parts with new ones (see Chapter 6).

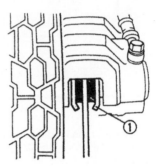

5.4 Look into the pad cavity from the rear of the caliper to inspect the rear pads (FZ600 shown; FJ600 and XJ600 similar)

6.1 If the rear brake shoe indicator pointer moves past the end of the "safe movement" area on the scale, the shoes are worn and should be replaced

3   Make sure all brake fasteners are tight. Check the brake pads (and rear shoes if equipped) for wear (see Section 5) and make sure the fluid level in the reservoir(s) is correct (see Section 3). Look for leaks at the hose connections and check for cracks in the hoses. If the lever is

7.7 Hold the switch body and turn the plastic nut (arrow) to adjust switch position

8.2 To adjust the front brake lever free play, loosen the locknut (arrow) and turn the screw; be sure to tighten the locknut after adjustment

spongy (or if the pedal is spongy on models with rear disc brakes), bleed the brakes as described in Chapter 6.

4   Make sure the brake light operates when the brake lever or pedal is depressed.

5   Make sure the brake light is activated just before the rear brake takes effect.

6   If adjustment is necessary, remove the following components for access:

  a) *XJ/FJ600 - if necessary, remove the right side cover.*
  a) *FZ600 - remove two bolts and take off the protective plate.*
  b) *YX600 Radian - remove the carburetor cover (Chapter 3) and the right side cover (Chapter 7).*

7   Hold the switch and turn the adjusting nut on the switch body **(see illustration)** until the brake light is activated when required. If the switch doesn't operate the brake lights, check it as described in Chapter 8.

8   The front brake light switch is not adjustable. If it fails to operate properly, replace it with a new one (see Chapter 8).

8   Brake lever play and pedal position - check and adjustment

Front brake

Refer to illustration 8.2

1   The front brake lever must have the amount of free play listed in this Chapter's Specifications to prevent brake drag.

2   Operate the lever and check free play. If it's not correct, loosen the adjuster locknut, turn the adjuster to bring free play within the Specifications and tighten the locknut **(see illustration)**.

Rear brake

3   The rear brake pedal on all models should be positioned below the top of the footpeg the distance listed in this chapter's Specifications. On drum brake models, brake free play should be adjusted manually; on disc brake models, free play is adjusted automatically.

Drum brake models

Refer to illustration 8.5

4   Brake pedal height is adjusted by a bolt that presses against the brake pedal arm. To adjust the position of the pedal, loosen the locknut on the adjuster bolt, turn the adjuster bolt to set the pedal position and tighten the locknut.

5   Check pedal free play and compare it to the value listed in this Chapter's Specifications. If it's incorrect, turn the adjuster on the rear brake lever to change it **(see illustration)**. If adjustment causes the brake wear indicator to go past the end of the scale **(see illustration 6.1)**, disassemble the brake and check for wear (see Chapter 6).

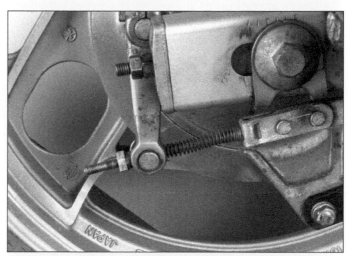

8.5 To adjust brake pedal free play on drum brake models, pull the brake lever forward away from the adjusting nut and turn the adjusting nut

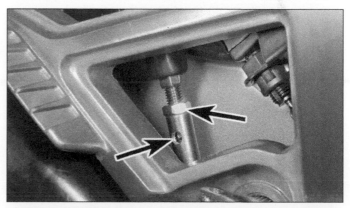

8.6 To adjust the brake pedal height on disc brake models, loosen the locknut (upper arrow) and turn the adjusting bolt; the end of the adjusting bolt must be visible in the hole (lower arrow) after adjustment

Disc brake models

Refer to illustration 8.6

6   Check pedal height and compare it to the value listed in this Chapter's Specifications. Adjust it if necessary by loosening the locknuts on the adjusting bolt, turning the adjusting bolt as needed and tightening the locknuts **(see illustration)**.

9.4 Use an accurate gauge to check the air pressure in the tires

10.2 Twist the throttle grip lightly and check freeplay

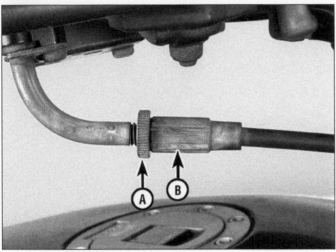

10.6 To adjust throttle freeplay, loosen the lock wheel (A), turn the adjuster (B), then tighten the lock wheel

7   Check pedal free play and compare it to the value listed in this Chapter's Specifications. If it's incorrect, check the rear brake for worn pads (Section 5). If the pads are okay, check the brake for wear or damage (see Chapter 6).

All models

8   If necessary, adjust the brake light switch (see Section 7).

---

9   Tires/wheels - general check

Refer to illustration 9.4

1   Routine tire and wheel checks should be made with the realization that your safety depends to a great extent on their condition.
2   Check the tires carefully for cuts, tears, embedded nails or other sharp objects and excessive wear. Operation of the motorcycle with excessively worn tires is extremely hazardous, as traction and handling are directly affected. Measure the tread depth at the center of the tire and replace worn tires with new ones when the tread depth is less than specified.
3   Repair or replace punctured tires as soon as damage is noted. Do not try to patch a torn tire, as wheel balance and tire reliability may be impaired.

4   Check the tire pressures when the tires are cold and keep them properly inflated **(see illustration)**. Proper air pressure will increase tire life and provide maximum stability and ride comfort. Keep in mind that low tire pressures may cause the tire to slip on the rim or come off, while high tire pressures will cause abnormal tread wear and unsafe handling.
5   The cast wheels used on these machines are virtually maintenance free, but they should be kept clean and checked periodically for cracks and other damage. Never attempt to repair damaged cast wheels; they must be replaced with new ones.
6   Check the valve stem locknuts to make sure they are tight. Also, make sure the valve stem cap is in place and tight. If it is missing, install a new one made of metal or hard plastic.

---

10   Throttle operation/grip freeplay - check and adjustment

Throttle check

Refer to illustration 10.2

1   Make sure the throttle grip rotates easily from fully closed to fully open with the front wheel turned at various angles. The grip should return automatically from fully open to fully closed when released. If the throttle sticks, check the throttle cables for cracks or kinks in the housings. Also, make sure the inner cables are clean and well-lubricated.
2   Check for a small amount of free play at the grip and compare the freeplay to the value listed in this chapter's Specifications **(see illustration)**. If adjustment is necessary, adjust idle speed first (see Section 19).

Throttle cable adjustment

Refer to illustration 10.6

3   Freeplay adjustments are made at the carburetor end of the cable.
4   Make sure the throttle grip is in the fully closed position.
5   Make sure the throttle linkage lever contacts the idle adjusting screw when the throttle grip is in the closed throttle position.
6   At the throttle grip, loosen the lock wheel **(see illustration)**. Turn the adjuster to achieve the correct freeplay, then tighten the locknut. **Warning**: *Turn the handlebars all the way through their travel with the engine idling. Idle speed should not change. If it does, the cables may be routed incorrectly. Correct this dangerous condition before riding the bike.*

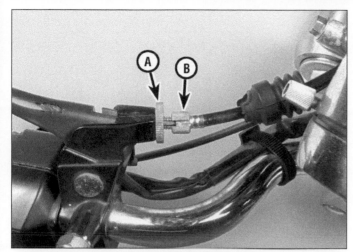

11.2 Pull back the rubber boot to adjust clutch lever position; loosen the lock wheel (A) and turn the adjuster (B), then tighten the lock wheel

11.3 To adjust the clutch cable at the engine, loosen the nuts (arrows), turn them to move the cable housing, then tighten the nuts

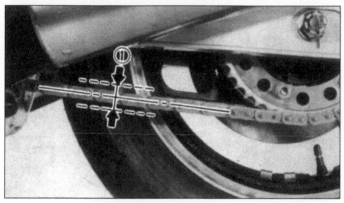

12.3 Push up on the lower run of the chain and measure how far it deflects - if it's not within the specified limits, adjust the slack in the chain

1   Chain deflection

12.4a The chain guard is secured by screws at the front (arrow) . . .

11 Clutch - check

Refer to illustrations 11.2 and 11.3

1   The clutch release mechanism is adjustable at the lever and engine attachment locations.
2   Start at the lever. Pull the rubber cover back, loosen the lock wheel and turn the adjuster in or out as required **(see illustration)**.
3   If you can't adjust freeplay to specifications with the handlebar adjuster, loosen the two nuts on the cable and adjust the housing position. Relock the two nuts to secure the housing into its final position **(see illustration)**.
4   Start the bike, release the clutch and ride off, noting the position of the clutch lever when the clutch begins to engage. If it's too close to the handlebar with freeplay correctly adjusted, check the cable and clutch components for wear and damage (see Chapter 2).

12 Drive chain and sprockets - check, adjustment and lubrication

Check

Refer to illustrations 12.3, 12.4a, 12.4b, 12.4c and 12.5

1   A neglected drive chain won't last long and can quickly damage the sprockets. Routine chain adjustment and lubrication isn't difficult and will ensure maximum chain and sprocket life.
2   To check the chain, place the bike on its centerstand (if equipped) or prop it securely in an upright position. Shift the transmission into Neutral. Make sure the ignition switch is Off.
3   Push up on the bottom run of the chain and measure the slack midway between the two sprockets **(see illustration)**, then compare your measurements to the value listed in this chapter's Specifications. As wear occurs, the chain will actually stretch, which means adjustment usually involves removing some slack from the chain. In some cases where lubrication has been neglected, corrosion and galling may cause the links to bind and kink, which effectively shortens the chain's length. If the chain is tight between the sprockets, rusty or kinked, it's time to replace it with a new one. **Note**: *Repeat the chain slack measurement along the length of the chain - ideally, every inch or so. If you find a tight area, mark it with felt pen or paint and repeat the measurement after the bike has been ridden. If the chain is still tight in the same area, it may be damaged or worn. Because a tight or kinked chain can damage the transmission output shaft bearing, it's a good idea to replace it.*
4   Remove the chain guard **(see illustrations)**. Check the entire length of the chain for damaged rollers, loose links and pins. Pull the

12.4b . . . and at the rear (arrow) (YX600 shown)

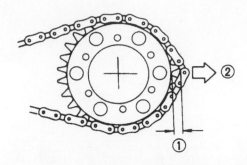

12.4c If the chain can be pulled more than half the length of a tooth away from the sprocket, it's worn and should be replaced

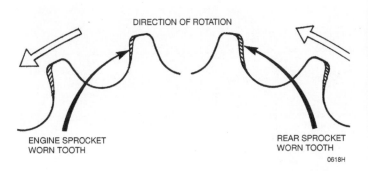

12.5 Check the sprockets in the areas indicated to see if they're worn excessively

12.7 Remove the cotter pin and loosen the axle nut

chain rearward, away from the center of the rear sprocket (**see illustration**). If the chain pulls away by more than half the length of a sprocket tooth, it's worn and should be replaced. Rotate the wheel and repeat this check at several places on the chain, since it may wear unevenly. **Note:** *Never install a new chain on old sprockets, and never use the old chain if you install new sprockets - replace the chain and sprockets as a set.*

5   Remove the engine sprocket cover (see Chapter 5). Check the teeth on the engine sprocket and the rear sprocket for wear (**see illustration**).

Adjustment

Refer to illustrations 12.7, 12.8, 12.10a and 12.10b

6   Rotate the rear wheel until the chain is positioned with the least amount of slack present.

7   Remove the cotter pin from the axle nut and loosen the nut (**see illustration**).

8   Loosen and back-off the locknuts on the adjuster bolts (**see illustration**).

9   Turn the axle adjusting bolts on both sides of the swingarm until the proper chain tension is obtained (get the adjuster on the chain side close, then set the adjuster on the opposite side). Be sure to turn the adjusting bolts evenly to keep the rear wheel in alignment. If the adjusting bolts reach the end of their travel, the chain is excessively worn and should be replaced with a new one (see Chapter 6).

10   When the chain has the correct amount of slack, make sure the marks on the adjusters correspond to the same relative marks on each

12.8 Loosen the locknut (arrow) on each chain adjuster so the adjuster can be turned

side of the swingarm (**see illustration**). Tighten the axle nut to the torque listed in the Chapter 6 Specifications, then install a new cotter pin and bend it properly (**see illustration**). If necessary, turn the nut an additional amount to line up the cotter pin hole with the castellations in the nut - don't loosen the nut to do this.

11   Tighten the chain adjuster locknuts securely.

12.10a When the adjuster bolts are set evenly, the adjuster marks on both sides should line up with the same marks in the swingarm, but don't rely completely on this; make a visual check of sprocket alignment as well

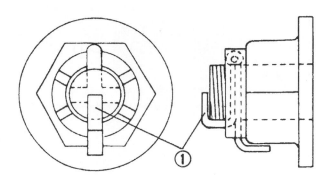

12.10b This is the correct way to bend the axle nut cotter pin - always use a new cotter pin whenever the old one is removed

1   Cotter pin

13.6 Remove the oil pan drain plug (on FZ600 models, it's on the left side of the engine)

13.8 Remove the engine drain plug (1) and the filter bolt (2)

Lubrication

Note: *If the chain is extremely dirty, it should be removed and cleaned before it's lubricated (see Chapter 5).*

12   The best time to lubricate the chain is after the motorcycle has been ridden. When the chain is warm, the lubricant will penetrate the joints between the side plates, pins, bushings and rollers to provide lubrication of the internal load bearing areas. **Note**: *If the chain is dirty, remove and clean it (see Chapter 5). Yamaha specifies SAE 30 to SAE 50 engine oil only; do not use chain lube, which may contain solvents that could damage the O-rings. Apply the oil to the area where the side plates overlap - not the middle of the rollers. Apply the oil to the top of the lower chain run, so centrifugal force will work the oil into the chain when the bike is moving. After applying the lubricant, let it soak in a few minutes before wiping off any excess.*

13 Engine oil/filter - change

Refer to illustrations 13.6, 13.8, 13.9a and 13.9b

1   Consistent routine oil and filter changes are the single most important maintenance procedure you can perform on a motorcycle. The oil not only lubricates the internal parts of the engine, transmission

and clutch, but it also acts as a coolant, a cleaner, a sealant, and a protectant. Because of these demands, the oil takes a terrific amount of abuse and should be replaced often with new oil of the recommended grade and type. Saving a little money on the difference in cost between a good oil and a cheap oil won't pay off if the engine is damaged.

2   Before changing the oil and filter, warm up the engine so the oil will drain easily. Be careful when draining the oil, as the exhaust pipes, the engine, and the oil itself can cause severe burns.

3   Put the motorcycle on the centerstand (if equipped) or prop it securely upright. Place a clean drain pan beneath the crankcase drain plug and filter housing. Remove the oil filler cap to vent the crankcase and act as a reminder that there is no oil in the engine.

4   If you're working on an XJ600 or FJ600, refer to Chapter 7 and remove the lower fairing.

5   If you're working on an FZ600, remove the exhaust system and center and lower fairings (see Chapters 3 and 7).

6   Next, remove the drain plug from the engine **(see illustration)** and allow the oil to drain into the pan. Discard the sealing washer on the drain plug; it should be replaced whenever the plug is removed.

7   Make sure the oil drain pan is under the oil filter, then loosen the filter bolt three or four turns. If additional maintenance is planned for this time period, check or service another component while the oil is allowed to drain completely.

8   Remove the oil filter bolt and take off the filter cover **(see illustration)**. **Note**: *Oil will run out of the filter cover onto the frame rails. There's no way to prevent this, so have some rags handy to wipe up the spilled oil.*

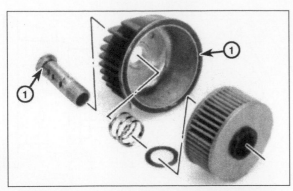

13.9a Oil filter components; be sure the O-rings (1) are installed correctly

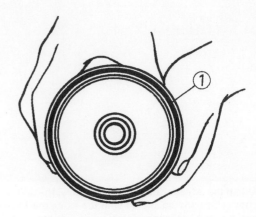

13.9b The large O-ring (1) fits into the filter housing groove

14.3 Remove the cover screws and lift off the cover . . .

16   If there are pieces of fiber-like material in the oil, the clutch is wearing excessively and should be checked.

17   Add the recommended type and amount of oil and install the filler cap. With the engine idling, slightly loosen the oil gallery bolt at the top right of the cylinder head. If oil doesn't seep from the threads within one minute, shut the engine off and find out why. If it does seep, retighten the bolt. Check for oil leaks, then shut the engine off and recheck oil level.

18   The old oil drained from the engine cannot be reused in its present state and should be disposed of. Check with your local refuse disposal company, disposal facility or environmental agency to see whether they will accept the used oil for recycling. Don't pour used oil into drains or onto the ground. After the oil has cooled, it can be drained into a suitable container (capped plastic jugs, topped bottles, milk cartons, etc.) for transport to one of these disposal sites.

---

14   Air filter element - servicing

Refer to illustrations 14.3 and 14.4

FJ600, XJ600 and YX Radian models

1   If you're working on an FJ600, XJ600 or YX600 Radian, remove the left side cover (see Chapter 7).

2   If you're working on a Radian, remove the carburetor cover as well.

3   Remove the cover screws and lift off the housing cover **(see illustration)**. Inspect the cover seal and replace it if it's damaged or deteriorated.

4   Lift off the housing cover and remove the element **(see illustration)**. Wipe out the housing with a clean rag.

FZ600 models

5   Remove the seat and side covers (see Chapter 7).

6   Remove the fuel tank (see Chapter 3).

7   Remove the cover from the top of the air filter case and lift the element out.

All models

8   Tap the element on a hard surface to shake out dirt. If compressed air is available, use it to clean the element by blowing from the inside out. If the element is extremely dirty or torn, replace the element with a new one.

9   Remove the filter element, washer and spring from the cover and engine **(see illustration)**. Remove the O-ring from its slot in the cover **(see illustration)**.

10   Clean the filter cover and housing with solvent or clean rags. Make sure the holes in the filter bolt are clear. Wipe any remaining oil off the filter sealing area of the crankcase.

11   Clean the components and check them for damage - especially, be sure to check the spring for distortion. If any damage is found, replace the damaged part(s).

12   Check the condition of the drain plug threads and the sealing washer.

13   Install a new O-ring in the cover groove and make sure it's positioned securely **(see illustration 13.9b)**. Install the spring, washer and filter element in the cover, install the cover on the engine and tighten the bolt to the torque listed in this Chapter's Specifications.

14   Slip a new sealing washer over the crankcase drain plug, then install and tighten them to the torque listed in this Chapter's Specifications. Avoid overtightening, as damage to the engine case will result.

15   Before refilling the engine, check the old oil carefully. If the oil was drained into a clean pan, small pieces of metal or other material can be easily detected. If the oil is very metallic colored, then the engine is experiencing wear from break-in (new engine) or from insufficient lubrication. If there are flakes or chips of metal in the oil, then something is drastically wrong internally and the engine will have to be disassembled for inspection and repair.

14.4 . . . and remove the filter element from the air box

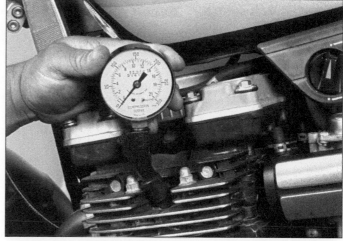

15.7 Check cylinder compression with a compression tester

9   Reinstall the filter by reversing the removal procedure. Make sure the element is seated properly in the filter housing before installing the cover.

10   Install all components removed for access.

15   Cylinder compression - check

Refer to illustration 15.7

1   Among other things, poor engine performance may be caused by leaking valves, incorrect valve clearances, a leaking head gasket, or worn pistons, rings and/or cylinder walls. A cylinder compression check will help pinpoint these conditions and can also indicate the presence of excessive carbon deposits in the cylinder heads.

2   The only tools required are a compression gauge and a spark plug wrench. Depending on the outcome of the initial test, a squirt-type oil can may also be needed.

3   Start the engine and allow it to reach normal operating temperature, then stop it.

4   Place the motorcycle on the centerstand (if equipped) or prop it securely in an upright position.

5   Remove the spark plugs (see Section 16, if necessary). Work carefully - don't strip the spark plug hole threads and don't burn your hands.

6   Disable the ignition by unplugging the primary wires from the coils (see Chapter 4). Be sure to mark the locations of the wires before detaching them.

7   Install the compression gauge in one of the spark plug holes **(see illustration)**. Hold or block the throttle wide open.

8   Crank the engine over a minimum of four or five revolutions (or until the gauge reading stops increasing) and observe the initial movement of the compression gauge needle as well as the final total gauge reading. Repeat the procedure for the other cylinders and compare the results to the value listed in this chapter's Specifications.

9   If the compression in all four cylinders built up quickly and evenly to the specified amount, you can assume the engine upper end is in reasonably good mechanical condition. Worn or sticking piston rings and worn cylinders will produce very little initial movement of the gauge needle, but compression will tend to build up gradually as the engine spins over. Valve and valve seat leakage, or head gasket leakage, is indicated by low initial compression which does not tend to build up.

10   To further confirm your findings, add a small amount of engine oil to each cylinder by inserting the nozzle of a squirt-type oil can through the spark plug holes. The oil will tend to seal the piston rings if they are leaking. Repeat the test for the other cylinders.

11   If the compression increases significantly after the addition of the oil, the piston rings and/or cylinders are definitely worn. If the

16.3a Rotate the spark plug caps back and forth to loosen them, then pull them off the plugs and check them for brittleness and cracking

compression does not increase, the pressure is leaking past the valves or the head gasket. Leakage past the valves may be due to insufficient valve clearances, burned, warped or cracked valves or valve seats or valves that are hanging up in the guides.

12   If compression readings are considerably higher than specified, the combustion chambers are probably coated with excessive carbon deposits. It is possible (but not very likely) for carbon deposits to raise the compression enough to compensate for the effects of leakage past rings or valves. Use of a fuel additive that will dissolve the adhesive bonding the carbon particles to the crown and chamber is the easiest way to remove the build-up. Otherwise, the cylinder head will have to be removed and decarbonized (Chapter 2).

16   Spark plugs - replacement

Refer to illustrations 16.3a, 16.3b, 16.7a and 16.7b

1   This motorcycle is equipped with spark plugs that have 12 mm threads and an 18 mm wrench hex. Make sure your spark plug socket is the correct size before attempting to remove the plugs.

2   If you're working on an FZ600, remove the center and lower fairing panels from the left and right sides of the bike (see Chapter 7).

3   Disconnect the spark plug caps from the spark plugs **(see illustration)**. If available, use compressed air to blow any accumulated

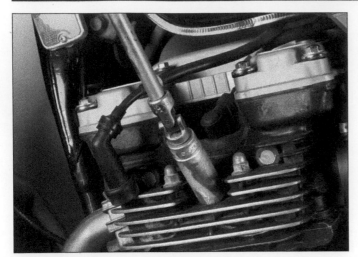

16.3b Use an extension and a deep socket (preferably one with a rubber insert to prevent damage to the plug) to remove the spark plugs

16.7a Spark plug manufacturers recommend using a wire type gauge when checking the gap - if the wire doesn't slide between the electrodes with a slight drag, adjustment is required

debris from around the spark plugs. Remove the plugs **(see illustration)**.

4   Inspect the electrodes for wear. Both the center and side electrodes should have square edges and the side electrode should be of uniform thickness. Look for excessive deposits and evidence of a cracked or chipped insulator around the center electrode. Compare your spark plugs to the spark plug reading chart. Check the threads, the washer and the ceramic insulator body for cracks and other damage.

5   If the electrodes are not excessively worn, and if the deposits can be easily removed with a wire brush, the plugs can be regapped and reused (if no cracks or chips are visible in the insulator). If in doubt concerning the condition of the plugs, replace them with new ones, as the expense is minimal.

6   Cleaning spark plugs by sandblasting is permitted, provided you clean the plugs with a high flash-point solvent afterwards.

7   Before installing new plugs, make sure they are the correct type and heat range. Check the gap between the electrodes, as they are not preset. For best results, use a wire-type gauge rather than a flat gauge to check the gap **(see illustration)**. If the gap must be adjusted, bend the side electrode only and be very careful not to chip or crack the insulator nose **(see illustration)**. Make sure the washer is in place before installing each plug.

8   Since the cylinder head is made of aluminum, which is soft and easily damaged, thread the plugs into the heads by hand. Since the plugs are recessed, slip a short length of hose over the end of the plug to use as a tool to thread it into place. The hose will grip the plug well enough to turn it, but will start to slip if the plug begins to cross-thread in the hole - this will prevent damaged threads and the resultant repair costs.

9   Once the plugs are finger tight, the job can be finished with a socket. If a torque wrench is available, tighten the spark plugs to the torque listed in this chapter's Specifications. If you do not have a torque wrench, tighten the plugs finger tight (until the washers bottom on the cylinder head) then use a wrench to tighten them an additional 1/4 turn. Regardless of the method used, do not over-tighten them.

10   Reconnect the spark plug caps and reinstall all components removed for access.

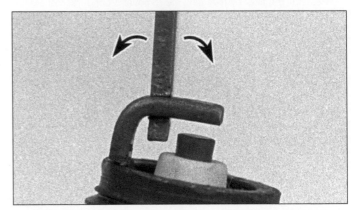

16.7b To change the gap, bend the side electrode only, as indicated by the arrows, and be very careful not to crack or chip the ceramic insulator surrounding the center electrode

17.2a Lubricate the front footpegs at their pivot points

17 Lubrication - general

Refer to illustrations 17.2a through 17.2f and 17.3

1   Since the controls, cables and various other components of a motorcycle are exposed to the elements, they should be lubricated

periodically to ensure safe and trouble-free operation.

2   The footpegs, clutch and brake levers, brake pedal, shift lever and side and centerstand (if equipped) pivots should be lubricated frequently **(see illustrations)**. In order for the lubricant to be applied where it will do the most good, the component should be disassembled. However, if chain and cable lubricant is being used, it can be applied to the pivot joint gaps and will usually work its way into the areas where friction occurs. If motor oil or light grease is being

Common spark plug conditions

NORMAL

Symptoms: Brown to grayish-tan color and slight electrode wear. Correct heat range for engine and operating conditions.
Recommendation: When new spark plugs are installed, replace with plugs of the same heat range.

WORN

Symptoms: Rounded electrodes with a small amount of deposits on the firing end. Normal color. Causes hard starting in damp or cold weather and poor fuel economy.
Recommendation: Plugs have been left in the engine too long. Replace with new plugs of the same heat range. Follow the recommended maintenance schedule.

CARBON DEPOSITS

Symptoms: Dry sooty deposits indicate a rich mixture or weak ignition. Causes misfiring, hard starting and hesitation.
Recommendation: Make sure the plug has the correct heat range. Check for a clogged air filter or problem in the fuel system or engine management system. Also check for ignition system problems.

ASH DEPOSITS

Symptoms: Light brown deposits encrusted on the side or center electrodes or both. Derived from oil and/or fuel additives. Excessive amounts may mask the spark, causing misfiring and hesitation during acceleration.
Recommendation: If excessive deposits accumulate over a short time or low mileage, install new valve guide seals to prevent seepage of oil into the combustion chambers. Also try changing gasoline brands.

OIL DEPOSITS

Symptoms: Oily coating caused by poor oil control. Oil is leaking past worn valve guides or piston rings into the combustion chamber. Causes hard starting, misfiring and hesitation.
Recommendation: Correct the mechanical condition with necessary repairs and install new plugs.

GAP BRIDGING

Symptoms: Combustion deposits lodge between the electrodes. Heavy deposits accumulate and bridge the electrode gap. The plug ceases to fire, resulting in a dead cylinder.
Recommendation: Locate the faulty plug and remove the deposits from between the electrodes.

TOO HOT

Symptoms: Blistered, white insulator, eroded electrode and absence of deposits. Results in shortened plug life.
Recommendation: Check for the correct plug heat range, over-advanced ignition timing, lean fuel mixture, intake manifold vacuum leaks, sticking valves and insufficient engine cooling.

PREIGNITION

Symptoms: Melted electrodes. Insulators are white, but may be dirty due to misfiring or flying debris in the combustion chamber. Can lead to engine damage.
Recommendation: Check for the correct plug heat range, over-advanced ignition timing, lean fuel mixture, insufficient engine cooling and lack of lubrication.

HIGH SPEED GLAZING

Symptoms: Insulator has yellowish, glazed appearance. Indicates that combustion chamber temperatures have risen suddenly during hard acceleration. Normal deposits melt to form a conductive coating. Causes misfiring at high speeds.
Recommendation: Install new plugs. Consider using a colder plug if driving habits warrant.

DETONATION

Symptoms: Insulators may be cracked or chipped. Improper gap setting techniques can also result in a fractured insulator tip. Can lead to piston damage.
Recommendation: Make sure the fuel anti-knock values meet engine requirements. Use care when setting the gaps on new plugs. Avoid lugging the engine.

MECHANICAL DAMAGE

Symptoms: May be caused by a foreign object in the combustion chamber or the piston striking an incorrect reach (too long) plug. Causes a dead cylinder and could result in piston damage.
Recommendation: Repair the mechanical damage. Remove the foreign object from the engine and/or install the correct reach plug.

17.2b Lubricate the rear footpegs at their pivot points

17.2c Lubricate the front brake lever at its pivot point, then lubricate the clutch lever in the same way

used, apply it sparingly as it may attract dirt (which could cause the controls to bind or wear at an accelerated rate). **Note**: *One of the best lubricants for the control lever pivots is a dry-film lubricant (available from many sources by different names).*

3   To lubricate the throttle cable, clutch cable and choke cables (if equipped), disconnect the cable at the lower end, then lubricate the cable with a pressure lube adapter **(see illustration)**. See Chapter 3 for the throttle and choke cable removal procedure and Chapter 2 for the clutch cable removal procedure. **Note**: *Yamaha recommends that the throttle grip be removed and lubricated whenever the throttle cables are lubricated. Refer to the handlebars section of Chapter 5.*

4   The speedometer cable should be removed from its housing and lubricated with motor oil or cable lubricant.

5   Refer to Chapter 5 for the swingarm needle bearing and rear suspension linkage (if equipped) lubrication procedures.

18   Valve clearances - check and adjustment

1   The engine must be completely cool for this maintenance procedure, so let the machine sit overnight before beginning.

2   Disconnect the cable from the negative terminal of the battery.

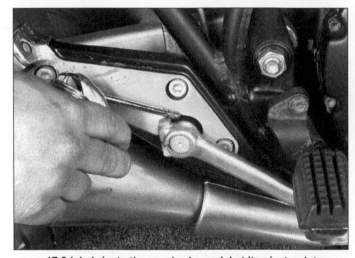

17.2d Lubricate the rear brake pedal at its pivot point

17.2e Lubricate the sidestand pivot point . . .

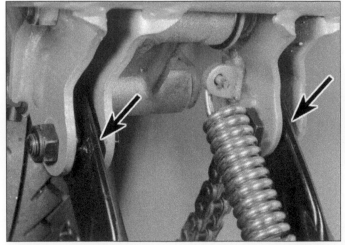

17.2f . . . and the centerstand pivot points (arrows) if equipped

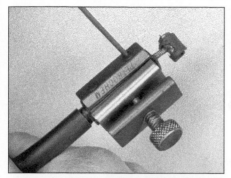

17.3 Lubricating a cable with a pressure lube adapter (make sure the tool seats around the inner cable)

18.8 On early models, remove four screws and detach the signal generator cover from the left side of the engine . . .

18.9 . . . and turn the crankshaft with a wrench on the turning bolt to align the T mark on the timing rotor with the line on the upper pickup coil

18.10 On later models, remove the Allen bolts (arrows) and remove the alternator cover . . .

18.11a . . . and turn the crankshaft with a socket on the bolt . . .

3 Remove the seat (see Chapter 7) and the fuel tank (see Chapter 3).

4 If you're working on an FJ600 or XJ600, remove the upper fairing and side covers (see Chapter 7).

5 If you're working on an FZ600, remove the side covers, center and lower fairings and air ducts (see Chapter 7).

6 If you're working on a YX600 Radian, remove the horn (see Chapter 8).

7 Remove the valve cover (see Chapter 2).

Early models

Refer to illustrations 18.8 and 18.9

8 Remove the signal generator cover to provide access to the crankshaft rotation bolt **(see illustration)**.

9 Position the no. 1 piston (at the left side of the engine) at Top Dead Center (TDC) on the compression stroke. Do this by turning the crankshaft, with an open end wrench placed on the flats of the timing rotor, until the T mark on the timing rotor aligns with the line on the upper pickup coil **(see illustration)**. **Note:** *Turn the crankshaft counterclockwise (anti-clockwise) viewed from the left side of the engine.*

Later models

Refer to illustrations 18.10, 18.11a and 18.11b

10 Remove the alternator cover to provide access to the crankshaft rotation bolt **(see illustration)**.

11 Position the no. 1 piston (on the left side of the engine) at Top Dead Center (TDC) on the compression stroke. Do this by turning the

18.11b . . . to align the T mark on the alternator rotor with the mark on the crankcase

crankshaft, with a socket placed on the alternator rotor bolt, until the T mark on the rotor is aligned with the stationary pointer on the pickup coil **(see illustrations)**. **Note:** *Turn the crankshaft counterclockwise (anti-clockwise) viewed from the left side of the engine.*

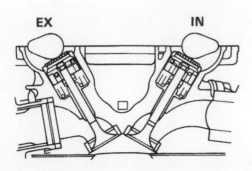

18.12 Make sure the cam lobes for no. 1 cylinder point away from each other as shown (the no. 4 cylinder cam lobes will point toward each other) - if the no. 1 cam lobes are pointing toward each other, no. 1 cylinder is at TDC on the exhaust stroke; rotate the engine one full turn to bring no. 1 cylinder to TDC compression

18.13 Slip a feeler gauge between the cam lobe and valve adjusting shim - it should pull out with a very light drag

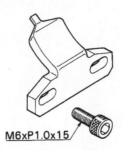

18.22 The valve lifter notch (arrow) should be toward the center of the engine

18.23a The Yamaha valve adjusting tool is attached to the cylinder head with a single Allen bolt to hold the lifter down for shim removal

All models

Refer to illustrations 18.12, 18.13 and 18.22

12   Now, check the position of the no. 1 cylinder cam lobes - they should be pointing away from each other **(see illustration)**. Piston no. 1 is now at TDC compression. If the cam lobes are pointing toward each other, no. 1 cylinder is at TDC exhaust. Turn the crankshaft one full turn.

13   Start with the no. 1 intake valve clearance (the intake valve is on the carburetor side of the engine). Insert a feeler gauge of the thickness listed in this chapter's Specifications between the cam lobe and valve adjusting shim **(see illustration)**. Pull the feeler gauge out slowly - you should feel a slight drag. If there's no drag, the clearance is too loose. If there's a heavy drag, the clearance is too tight.

14   If the clearance is incorrect, write down the actual measured clearance. You'll need this information later to select a new valve adjusting shim.

15   Now measure the no. 1 exhaust valve clearance, following the same procedure you used for the intake valve. Be sure to use a feeler gauge of the specified thickness and write down the actual clearances of any valves that aren't within the Specifications.

16   Rotate the crankshaft exactly one-half turn to place piston no. 2 at TDC compression. The cam lobes for no. 2 cylinder should now point away from each other **(see illustration 18.12)**.

17   Measure the clearances of both valves on cylinder no. 2 and write down any that aren't within the Specifications.

18   Rotate the crankshaft another half turn to place cylinder no. 4 at TDC (cam lobes pointing away from each other). Measure its valve clearances.

19   Rotate the engine another half turn to place cylinder no. 3 at TDC and measure its valve clearances.

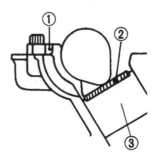

18.23b Turn the cam lobe so it presses the lifter down and install the adjusting tool

1 *Adjusting tool*	3 *Lifter*
2 *Adjusting shim*	

20   If any of the clearances need to be adjusted, go to Step 23 or Step 27.

21   If all the clearances were within the Specifications, go to Step 35.

22   Check the positions of the valve lifter notches. They should be turned so they're facing towards the spark plugs **(see illustration)**.

With Yamaha valve adjusting tool

Refer to illustrations 18.23a, 18.23b and 18.25

23   If you're using the Yamaha valve adjusting tool (US part no. YM-01245/UK part no. 90890-01245) **(see illustration)**, turn the camshaft until the cam lobe of the valve to be adjusted is pointing directly at the valve lifter **(see illustration)**.

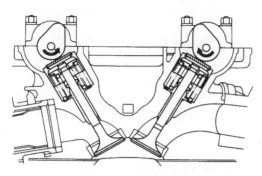

18.25 Turn the camshafts so the lobes are upward, but be sure to turn them only in the indicated directions so the lobe won't drag across the adjusting tool

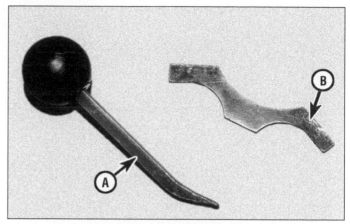

18.27 A universal adjusting tool can be used to adjust the valves; The wedge (A) is used to make a gap so the holder (B) can be installed between the camshaft and lifter - be sure the tool rests on the lifter and not on the adjusting shim

18.28 Push the wedge between the cam lobe and shim to create a gap, then insert the holder

18.29 Pry the shim loose from the lifter with the wedge

18.30 Lift the shim out with a magnet

18.31 There's a thickness mark on the underside of each valve shim (be sure to place the marked side of the shim downward during installation)

24   Install the tool on the valve lifter to hold it down. **Note:** *Be sure the lifter bears against the tool only, not the adjusting shim.*
25   Turn the camshaft so the lobe points upward **(see illustration)**. **Note:** *The camshaft must rotate inward (toward the center of the engine) so it doesn't drag on the holding tool. The tool will hold the valve lifter down, forming a gap so the shim can be removed.*
26   Slip a small screwdriver into the lifter notch, under the shim, and pry it up out of the lifter.

With universal valve adjusting tool

Refer to illustrations 18.27, 18.28 and 18.29
27   If you're using a universal valve adjusting tool **(see illustration)**, turn the camshaft so the lobe of the valve to be adjusted points upward **(see illustration 18.21)**.

28   Push the wedge of the valve adjusting tool between the cam lobe and adjusting shim so it pushes the lifter down slightly **(see illustration)**. Slip the holder between the camshaft and lifter, then pull out the wedge.
29   Slip the wedge tool between the adjusting shim and the lifter **(see illustration)**. This will dislodge the adjusting shim.

All models

Refer to illustrations 18.30, 18.31, 18.32a and 18.32b
30   Take the shim out with a magnet **(see illustration)**.
31   Determine the thickness of the shim you removed. It should be marked on the bottom of the shim **(see illustration)**, but the ideal way is to measure it with a micrometer. **Note**: *If the number on the shim*

[B] MEASURED CLEARANCE	200	205	210	215	220	225	230	235	240	245	250	255	260	265	270	275	280	285	290	295	300	305	310	315	320
0.00~0.05			200	205	210	215	220	225	230	235	240	245	250	255	260	265	270	275	280	285	290	295	300	305	310
0.06~0.10		200	205	210	215	220	225	230	235	240	245	250	255	260	265	270	275	280	285	290	295	300	305	310	315
0.11~0.15																									
0.16~0.20	205	210	215	220	225	230	235	240	245	250	255	260	265	270	275	280	285	290	295	300	305	310	315	320	
0.21~0.25	210	215	220	225	230	235	240	245	250	255	260	265	270	275	280	285	290	295	300	305	310	315	320		
0.26~0.30	215	220	225	230	235	240	245	250	255	260	265	270	275	280	285	290	295	300	305	310	315	320			
0.31~0.35	220	225	230	235	240	245	250	255	260	265	270	275	280	285	290	295	300	305	310	315	320				
0.36~0.40	225	230	235	240	245	250	255	260	265	270	275	280	285	290	295	300	305	310	315	320					
0.41~0.45	230	235	240	245	250	255	260	265	270	275	280	285	290	295	300	305	310	315	320						
0.46~0.50	235	240	245	250	255	260	265	270	275	280	285	290	295	300	305	310	315	320							
0.51~0.55	240	245	250	255	260	265	270	275	280	285	290	295	300	305	310	315	320								
0.56~0.60	245	250	255	260	265	270	275	280	285	290	295	300	305	310	315	320									
0.61~0.65	250	255	260	265	270	275	280	285	290	295	300	305	310	315	320										
0.66~0.70	255	260	265	270	275	280	285	290	295	300	305	310	315	320											
0.71~0.75	260	265	270	275	280	285	290	295	300	305	310	315	320												
0.76~0.80	265	270	275	280	285	290	295	300	305	310	315	320													
0.81~0.85	270	275	280	285	290	295	300	305	310	315	320														
0.86~0.90	275	280	285	290	295	300	305	310	315	320															
0.91~0.95	280	285	290	295	300	305	310	315	320																
0.96~1.00	285	290	295	300	305	310	315	320																	
1.10~1.05	290	295	300	305	310	315	320																		
1.06~1.10	295	300	305	310	315	320																			
1.11~1.15	300	305	310	315	320																				
1 16~1.20	305	310	315	320																					
1.21~1.25	310	315	320																						
1.26~1.30	315	320																							
1.31~1.35	320																								

[A] INSTALLED PAD NUMBER

VALVE CLEARANCE (cold):
  0.11 ~ 0.15 mm (0.0043 ~ 0.0059 in)
Example: Installed is 250
         Measured clearance is 0.32 mm (0.013 in)
         Replace 250 pad with 270 pad
*Pad number: (example)
         Pad No. 250 = 2.50 mm (0.098 in)
         Pad No. 255 = 2.55 mm (0.100 in)
Always install pad with number down.

18.32a Intake valve shim selection chart

does not end in 0 or 5, round it off to the nearest zero or 5. For example, if the number on the shim is 258, round it off to 260. If it's 254, round it off to 255.

32   If the clearance (measured and written down earlier) was too large, you need a thicker shim. If the clearance was too small, you need a thinner shim. Calculate the thickness of the replacement shim by referring to the accompanying charts (see illustrations).

33   Install the new shim and recheck the clearance. If it's within the Specifications, the valve is properly adjusted.

34   Adjust any remaining valves that were not within the Specifications. Perform the adjustments in the same order as the cylinder firing order (1-2-4-3).

35   Install the valve cover.

36   Reconnect the negative cable to the battery.

37   Install all components removed for access.

19   Idle speed - check and adjustment

Refer to illustration 19.3

1   The idle speed should be checked and adjusted before and after the carburetors are synchronized and when it is obviously too high or too low. Before adjusting the idle speed, make sure the valve clearances and spark plug gaps are correct. Also, turn the handlebars back-and-forth and see if the idle speed changes as this is done. If it does, the throttle cable may not be adjusted correctly, or it may be worn out. This is a dangerous condition that can cause loss of control of the bike. Be sure to correct this problem before proceeding.

2   The engine should be at normal operating temperature, which is usually reached after 10 to 15 minutes of stop and go riding. Place the motorcycle on the centerstand (if equipped) or prop it securely upright and make sure the transmission is in Neutral.

3   With the engine idling, turn the throttle stop screw (see illustration) until the idle speed listed in this chapter's Specifications is obtained.

4   Snap the throttle open and shut a few times, then recheck the idle speed. If necessary, repeat the adjustment procedure.

5   If a smooth, steady idle can't be achieved, the fuel/air mixture may be incorrect. Refer to Chapter 3 for additional carburetor information.

20   Carburetor synchronization - check and adjustment

Refer to illustrations 20.11, 20.14, 20.15 and 20.16

Warning: Gasoline (petrol) is extremely flammable, so take extra precautions when you work on any part of the fuel system. Don't smoke or allow open flames or bare light bulbs near the work area, and don't work in a garage where a natural gas-type appliance (such as a water heater or clothes dryer) is present. If you spill any fuel on your skin, rinse it off immediately with soap and water. When you perform any kind of work on the fuel system, wear safety glasses and have a fire extinguisher suitable for a Class B type fire (flammable liquids) on hand.

1   Carburetor synchronization is simply the process of adjusting the carburetors so they pass the same amount of fuel/air mixture to each cylinder. This is done by measuring the vacuum produced in each cylinder. Carburetors that are out of synchronization will result in decreased fuel mileage, increased engine temperature, less than ideal throttle response and higher vibration levels.

2   To properly synchronize the carburetors, you will need some sort of vacuum gauge setup, preferably with a gauge for each cylinder, or a mercury manometer, which is a calibrated tube arrangement that utilizes columns of mercury to indicate engine vacuum.

3   A manometer can be purchased from a motorcycle dealer or accessory shop and should have the necessary rubber hoses supplied with it for hooking into the vacuum hose fittings on the carburetors.

B MEASURED CLEARANCE	**A** INSTALLED PAD NUMBER																								
	200	205	210	215	220	225	230	235	240	245	250	255	260	265	270	275	280	285	290	295	300	305	310	315	320
0.00~0.05				200	205	210	215	220	225	230	235	240	245	250	255	260	265	270	275	280	285	290	295	300	305
0.06~0.10			200	205	210	215	220	225	230	235	240	245	250	255	260	265	270	275	280	285	290	295	300	305	310
0.11~0.15		200	205	210	215	220	225	230	235	240	245	250	255	260	265	270	275	280	285	290	295	300	305	310	315
0.16~0.20																									
0.21~0.25	205	210	215	220	225	230	235	240	245	250	255	260	265	270	275	280	285	290	295	300	305	310	315	320	
0.26~0.30	210	215	220	225	230	235	240	245	250	255	260	265	270	275	280	285	290	295	300	305	310	315	320		
0.31~0.35	215	220	225	230	235	240	245	250	255	260	265	270	275	280	285	290	295	300	305	310	315	320			
0.36~0.40	220	225	230	235	240	245	250	255	260	265	270	275	280	285	290	295	300	305	310	315	320				
0.41~0.45	225	230	235	240	245	250	255	260	265	270	275	280	285	290	295	300	305	310	315	320					
0.46~0.50	230	235	240	245	250	255	260	265	270	275	280	285	290	295	300	305	310	315	320						
0.51~0.55	235	240	245	250	255	260	265	270	275	280	285	290	295	300	305	310	315	320							
0.56~0.60	240	245	250	255	260	265	270	275	280	285	290	295	300	305	310	315	320								
0.61~0.65	245	250	255	260	265	270	275	280	285	290	295	300	305	310	315	320									
0.66~0.70	250	255	260	265	270	275	280	285	290	295	300	305	310	315	320										
0.71~0.75	255	260	265	270	275	280	285	290	295	300	305	310	315	320											
0.76~0.80	260	265	270	275	280	285	290	295	300	305	310	315	320												
0.81~0.85	265	270	275	280	285	290	295	300	305	310	315	320													
0.86~0.90	270	275	280	285	290	295	300	305	310	315	320														
0.91~0.95	275	280	285	290	295	300	305	310	315	320															
0.96~1.00	280	285	290	295	300	305	310	315	320																
1.10~1.05	285	290	295	300	305	310	315	320																	
1.06~1.10	290	295	300	305	310	315	320																		
1.11~1.15	295	300	305	310	315	320																			
1.16~1.20	300	305	310	315	320																				
1.21~1.25	305	310	315	320																					
1.26~1.30	310	315	320																						
1.31~1.35	315	320																							
1.36~1.40	320																								

VALVE CLEARANCE (cold):
  0.16 ~ 0.20 mm (0.0062 ~ 0.0079 in)
Example: Installed is 250
        Measured clearance is 0.32 mm (0.013 in)
        Replace 250 pad with 265 pad
*Pad number: (example)
        Pad No. 250 = 2.50 mm (0.098 in)
        Pad No. 255 = 2.55 mm (0.100 in)
Always install pad with number down.

18.32b Exhaust valve shim selection chart

19.3 Turn the throttle stop screw (arrow) in or out until the correct idle speed is obtained (carburetors removed for clarity)

20.11 Remove the vacuum hoses or fitting caps so a test gauge setup can be installed

4   A vacuum gauge setup can also be purchased from a dealer or fabricated from commonly available hardware and automotive vacuum gauges.

5   The manometer is the more reliable and accurate instrument, and for that reason is preferred over the vacuum gauge setup; however, since the mercury used in the manometer is a liquid, and extremely toxic, extra precautions must be taken during use and storage of the instrument.

6   Because of the nature of the synchronization procedure and the need for special instruments, most owners leave the task to a dealer service department or a reputable motorcycle repair shop.

7   If you're working on an XJ600 or an FJ600, remove the seat and side covers (see Chapter 7). If you're working on a 1984 or 1985 model, remove the fuel tank mounting bolt so you can lift the tank for access. If you're working on a 1989 or later XJ600, remove the fuel tank (see Chapter 3) and install an auxiliary fuel tank.

8   If you're working on an FZ600, remove the seat and center and lower fairings (see Chapter 7). Remove the fuel tank (see Chapter 3) and install an auxiliary fuel tank.

9   If you're working on a YX600 Radian, remove the seat and side covers (see Chapter 7). Remove the fuel tank mounting bolt so you can lift the rear of the tank for access.

10   Start the engine and let it run until it reaches normal operating temperature, then shut it off.

11   Detach the hoses or caps from the vacuum fittings on the carburetors **(see illustration)**, then hook up the vacuum gauge set or

20.14 Turn this screw (arrow) to synchronize carburetors no. 1 and 2 to each other (it's located between the two left-hand carburetors)

20.15 Turn this screw (arrow) to synchronize carburetors no. 3 and 4 to each other (it's located between the two right-hand carburetors)

20.16 Turn this screw (arrow) to synchronize the two pairs of carburetors to each other (it's located in the center of the carburetor assembly)

the manometer according to the manufacturer's instructions. Make sure there are no leaks in the setup, as false readings will result.

12   Start the engine and make sure the idle speed is correct. If it isn't, adjust it (see Section 19).

13   The vacuum readings for all of the cylinders should be the same, or at least within the tolerance listed in this chapter's Specifications. If the vacuum readings vary, adjust as necessary.

14   To perform the adjustment, synchronize the carburetors for no. 1 and no. 2 cylinders by turning the synchronizing screw, as needed, until the vacuum is identical or nearly identical for both cylinders **(see illustration)**. Snap the throttle open and shut 2 or 3 times, then recheck the adjustment and readjust as necessary.

15   Next synchronize the carburetors for no. 3 and no. 4 cylinders to each other by turning the synchronizing screw for those two carburetors **(see illustration)**. As with the no. 1 and no. 2 carburetors, snap the throttle open and shut 2 or 3 times, then recheck the adjustment and readjust as necessary.

16   Finally, turn the center synchronizing screw to synchronize the two pairs of carburetors to each other **(see illustration)**.

17   When the adjustment is complete, recheck the vacuum readings and idle speed, then stop the engine. Remove the vacuum gauge or manometer and attach the hoses or caps to the fittings on the carburetors.

21 Evaporative emission control system (California models only) - check

Refer to illustration 21.2

1   This system, installed on California models to conform to stringent emission control standards, routes fuel vapors from the fuel system into the engine to be burned, instead of letting them evaporate into the atmosphere. When the engine isn't running, vapors are stored in a carbon canister.

2   To begin the inspection of the system, remove the seat (Chapter 7) and fuel tank (Chapter 3). If you're working on an FZ600, remove the side covers and the center and lower fairings. Inspect the hoses from the fuel tank and carburetors to the canister for cracking, kinks or other signs of deterioration **(see illustration)**.

3   Label and disconnect the hoses, then remove the canisters from the machine.

4   Inspect the canisters for cracks or other signs of damage. Tip each canister so the nozzle points down. If fuel runs out of the canister, it is probably damaged internally, so it would be a good idea to replace it.

22 Exhaust system - check

1   Periodically check all of the exhaust system joints for leaks and loose fasteners. The lower fairing(s) (if equipped) will have to be removed to do this properly (see Chapter 7). If tightening the clamp bolts fails to stop any leaks, replace the gaskets with new ones (a procedure which requires disassembly of the system).

2   The exhaust pipe flange nuts at the cylinder heads are especially prone to loosening, which could cause damage to the head. Check them frequently and keep them tight.

23 Steering head bearings - check, adjustment and lubrication

1   This vehicle is equipped with ball steering head bearings which can become dented, rough or loose during normal use of the machine. In extreme cases, worn or loose steering head bearings can cause steering wobble that is potentially dangerous.

Check

Refer to illustration 23.4

2   To check the bearings, place the motorcycle on the centerstand and block the machine so the front wheel is in the air.

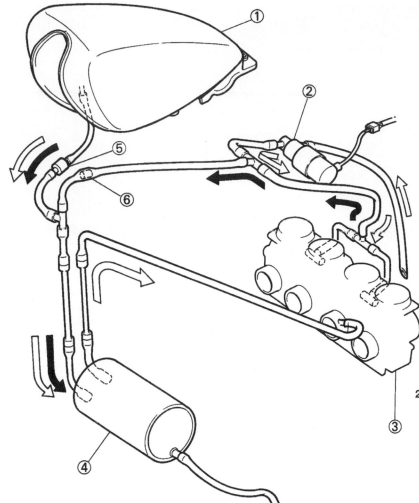

21.2 Evaporative emission control system details
(YX600 Radian shown; others similar)

1   Fuel tank
2   Electric solenoid
3   Carburetors
4   Canister
5   Rollover valve
6   Line restrictor

23.4 Grasp the front wheel and try to pull it back and forth; if it
moves, the steering head bearings are loose and in need
of adjustment

23.5 Loosen the lower triple clamp bolts on each fork (arrows)

3   Point the wheel straight ahead and slowly move the handlebars
from side-to-side. Dents or roughness in the bearing races will be felt
and the bars will not move smoothly.
4   Next, grasp the wheel and try to move it forward and backward
(see illustration). Any looseness in the steering head bearings will be
felt as front-to-rear movement of the fork legs. If play is felt in the
bearings, adjust the steering head as follows:

Adjustment

Refer to illustrations 23.5 and 23.7

5   Loosen the lower triple clamp bolts (see illustration). This allows
the necessary vertical movement of the steering stem in relation to the
fork tubes.
6   Remove the handlebars and upper triple clamp (see Section 27
and Chapter 5).

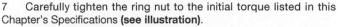

23.7 The ring nut can be tightened with a spanner wrench (C-spanner) like this one, but ideally, the spanner wrench should have a hole in the handle so a torque wrench can be used with it

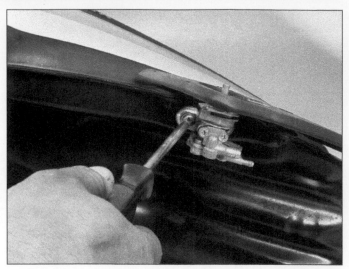

25.8 Fuel tap and filter can be removed with a screwdriver

7   Carefully tighten the ring nut to the initial torque listed in this Chapter's Specifications **(see illustration)**.
8   Turn the steering from lock to lock and check for binding. If there is any, slowly loosen the ring nut while you continue to turn the steering from lock to lock. The steering should turn smoothly without loosening it enough to create play. If this isn't possible, remove the bearings for inspection (see Chapter 5).
9   If the steering operates properly, proceed to the next step. If there is evidence of binding or rubbing, stop and investigate why.
10   Install the upper triple clamp and handlebars.
11   Recheck the steering head bearings for play as described in Step 4 above. The handlebars should move from a full turn to the full opposite turn with just a tap on the end of the handlebar. If necessary, repeat the adjustment procedure. Reinstall all parts previously removed. Tighten the steering stem nut, triple clamp bolts and handlebar bolts to the torques listed in the Chapter 5 Specifications.

Lubrication

12   Periodic cleaning and repacking of the steering head bearings is recommended by the manufacturer. Refer to Chapter 5 for steering head bearing lubrication and replacement procedures.

24 Fasteners - check

1   Since vibration of the machine tends to loosen fasteners, all nuts, bolts, screws, etc. should be periodically checked for proper tightness.
2   Pay particular attention to the following:

Spark plugs
Engine oil drain plug
Oil filter cover bolt and drain plug
Gearshift lever
Footpegs, sidestand and centerstand (if equipped)
Engine mount bolts
Shock absorber mount bolts
Rear suspension linkage bolts (if equipped)
Front axle and clamp bolt
Rear axle nut

3   If a torque wrench is available, use it along with the torque specifications at the beginning of this, or other, Chapters.
4   Should you find certain fasteners that do come loose periodically, apply a high quality non-hardening liquid thread locking agent to the threads and reassembly to proper torque.

25 Fuel system - check and filter cleaning

Refer to illustration 25.8
Warning: *Gasoline (petrol) is extremely flammable, so take extra precautions when you work on any part of the fuel system. Don't smoke or allow open flames or bare light bulbs near the work area, and d work in a garage where a natural gas-type appliance (such as a water heater or clothes dryer) is present. If you spill any fuel on your skin, rinse it off immediately with soap and water. When you perform any kind of work on the fuel system, wear safety glasses and have a fire extinguisher suitable for a Class B type fire (flammable liquids) on hand.*
1   Check the fuel tank, the tank breather hose, the fuel tap, the lines and the carburetors for leaks and evidence of damage.
2   If carburetor gaskets are leaking, the carburetors should be disassembled and rebuilt by referring to Chapter 3.
3   If the fuel tap is leaking, tightening the screws may help. If leakage persists, the tap should be disassembled and repaired or replaced with a new one.
4   If the fuel lines are cracked or otherwise deteriorated, replace them with new ones.
5   Check the vacuum hose connected to the fuel tap. If it is cracked or otherwise damaged, replace it with a new one.
6   The fuel filter, which is attached to the fuel tap, may become clogged and should be removed and cleaned periodically. In order to clean the filter, the fuel tank must be drained and the fuel tap removed.
7   Remove the fuel tank (see Chapter 4). Drain the fuel into an approved fuel container.
8   Once the tank is emptied, loosen and remove the screws that attach the fuel tap to the tank **(see illustration)**. Remove the tap and filter.
9   Clean the filter with solvent and blow it dry with compressed air. If the filter is torn or otherwise damaged, replace the entire fuel tap with a new one. Check the mounting flange O-ring and the washers on the screws. If they are damaged, replace them with new ones.
10   Install the O-ring, filter and fuel tap on the tank, then install the tank. Refill the tank and check carefully for leaks around the mounting flange and screws.

26 Suspension - check

Refer to illustration 26.3
1   The suspension components must be maintained in top operating

26.3 Check above and below the fork seals (arrow) for signs of oil leakage

27.7 Remove the drain screw from each fork leg and be prepared for the oil to spill out; use new drain screw gaskets on installation

condition to ensure rider safety. Loose, worn or damaged suspension parts decrease the vehicle's stability and control.

2   While standing alongside the motorcycle, lock the front brake and push on the handlebars to compress the forks several times. See if they move up-and-down smoothly without binding. If binding is felt, the forks should be disassembled and inspected as described in Chapter 5.

3   Carefully inspect the area around the fork seals for any signs of fork oil leakage **(see illustration)**. If leakage is evident, the seals must be replaced as described in Chapter 5.

4   Check the tightness of all front suspension nuts and bolts to be sure none have worked loose.

5   Inspect the rear shock(s) for fluid leakage and tightness of the mounting fasteners. If leakage is found, the shock should be replaced.

6   Set the bike on its centerstand (if equipped) or support it securely upright with the rear wheel off the ground. Grab the swingarm on each side, just ahead of the axle. Rock the swingarm from side to side - there should be no discernible movement at the rear. If there's a little movement or a slight clicking can be heard, make sure the pivot shaft nuts are tight. If the pivot nuts are tight but movement is still noticeable, the swingarm will have to be removed and the bearings replaced as described in Chapter 5.

7   Inspect the tightness of the rear suspension nuts and bolts.

27 Fork oil - replacement

1   Move the machine onto the centerstand (if equipped) or prop it securely upright.

FX/XJ600 models

2   Remove the handlebars (see Chapter 5).

FZ600 models

3   Release the fork air pressure on both forks. **Warning:** *Release the air pressure slowly so oil won't spurt out.*

4   Remove the lower and center fairing panels and the fairing stays from under the motorcycle (see Chapter 7).

5   Remove the exhaust system (see Chapter 3).

All models

Refer to illustrations 27.7, 27.11, 27.12, 27.13 and 27.17

6   Place a drain pan under each fork.

7   Remove the fork drain screw from the bottom of each fork **(see illustration)**.

8   Carefully pump the front-end up and down to remove the remaining oil from the fork.

9   Return the drain screws to their respective holes. Tighten to the torque listed in this Chapter's Specifications.

27.11 Unscrew the fork cap bolt . . .

27.12 . . . and lift out the spring seat and spring

10   Place a block under the front wheel to support the tire.

11   Remove the cap at the top of the fork tube **(see illustration)**. The spring will be under a moderate amount of pressure, so be careful!

12   Remove the spring seat from the top of the spring **(see illustration)**.

27.13 Remove the spring; it will retain a little left-over oil when it's pulled out, so have a rag handy to catch the drips

28.6 Camchain tensioner locknut (A) and tensioner bolt (B)

13   Remove the spring **(see illustration)**.
14   Pour in the specified amount of approved fork oil. If fork oil level is specified for the bike you're working on, measure it and compare to the value listed in this Chapter's Specifications. Add or remove oil as necessary to achieve the correct level.
15   Slowly pump the forks up and down several times to distribute the oil.
16   Install the spring with its closer-wound coils at the top. Install the spring seat and cap.

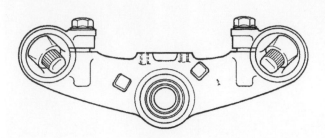

27.17 The air valves on FZ600 models must face inward and to the rear when the caps are installed

17   If you're working on an FZ600, face the air valve in the proper direction **(see illustration)**.
18   Tighten the cap to the torque listed in this Chapter's Specifications.
19   Install all parts removed for access.
20   If you're working on an FZ600, refer to Chapter 6 and adjust fork air pressure.

28 Cam chain adjustment

Refer to illustration 28.6
1   If you're working on an FZ600, remove the center and lower fairings (see Chapter 7).

Early models

2   Remove the signal generator cover from the left side of the engine **(see illustration 18.8)**.
3   Turn the crankshaft with an open end wrench on the flats of the timing rotor to align the C mark on the timing rotor with the line on the upper pickup coil.

Later models

4   Remove the alternator cover from the left side of the engine (see Chapter 8).
5   Turn the crankshaft with a socket on the alternator rotor bolt to align the C mark on the alternator rotor with the timing line on the crankcase.

All models

6   Loosen the locknut, then loosen the cam chain tensioner bolt **(see illustration)**. Tighten the bolt to the torque listed in the Chapter 2 Specifications, then tighten the locknut to the torque listed in the Chapter 2 Specifications.
7   Install all components removed for access.

Chapter 2
Engine, clutch and transmission

Contents

Specifications

General

Bore	58.5 mm 2.30 inches)
Stroke	55.7 mm (2.19 inches)
Displacement	599 cc
Compression ratio	10 to 1

Camshaft (FJ600, FZ600, YX600 Radian)

Lobe height (intake)	
Standard	36.25 to 36.35 mm(1.427 to 1.431 inches)
Minimum	36.2 mm (1.43 inches)
Lobe height (exhaust)	
Standard	35.75 to 35.85 mm (1.408 to 1.411 inches)
Minimum	35.7 mm (1.41 inches)
Base circle (intake)	
Standard	28.1 to 28.2 mm (1.106 to 1.110 inches)
Minimum	28.05 mm (1.1 inches)
Base circle (exhaust)	
Standard	28.05 to 28.15 mm (1.104 to 1.108 inches)
Minimum	28.0 mm (1.1 inches)
Bearing oil clearance	0.020 to 0.054 mm (0.0008 to 0.0021 inch)
Journal diameter	24.967 to 24.980 mm (0.9829 to 0.9834 inch)
Bearing bore	25.000 to 25.021 mm (0.9843 to 0.9850 inch)
Camshaft runout limit	
FJ600	0.03 mm (0.0012 inch)
FZ600, YX600 Radian	0.05 mm (0.002 inch)

Camshaft (XJ600)

Lobe height (intake)	
Standard	36.25 to 36.35 mm (1.427 to 1.431 inches)
Minimum	36.2 mm (1.43 inches)
Lobe height (exhaust)	
No. 1 and no. 4 cylinders	
Standard	35.75 to 35.85 mm (1.408 to 1.411 inches)
Minimum	35.7 mm (1.41 inches)
No. 2 and no. 3 cylinders	
Standard	32.65 to 36.35 mm (1.427 to 1.431 inches)
Minimum	36.2 mm (1.43 inches)
Base circle (intake)	
No. 1 and no. 4 cylinders	
Standard	28.1 to 28.2 mm (1.106 to 1.110 inches)
Minimum	28.05 mm (1.1 inches)
No. 2 and no. 3 cylinders	
Standard	28.075 to 28.175 mm (1.105 to 1.109 inches)
Minimum	28.025 mm (1.103 inches)
Base circle (exhaust)	
No. 1 and no. 4 cylinders	
Standard	28.05 to 28.15 mm (1.104 to 1.108 inches)
Minimum	28.0 mm (1.1 inches)
No. 2 and no. 3 cylinders	
Standard	28.03 to 28.13 mm (1.1035 to 1.107 inches)
Minimum	27.98 mm (1.101 inches)
Bearing oil clearance	0.020 to 0.054 mm (0.0008 to 0.0021 inch)
Journal diameter	24.967 to 24.980 mm (0.9829 to 0.9834 inch)
Bearing bore	25.000 to 25.021 mm (0.9843 to 0.9859 inch)
Camshaft runout limit	0.05 mm (0.002 inch)

Cylinder head, valves and valve springs

Cylinder head warpage limit	0.03 mm (0.0012 inch)
Valve stem bend limit	0.01 mm (0.0004 inch)
Valve head diameter	
Intake	31.4 to 31.6 mm (1.236 to 1.244 inches)
Exhaust	26.9 to 27.1 mm (1.059 to 1.067 inches)
Valve stem diameter	
Standard	
Intake	5.975 to 5.990 mm (0.235 to 0.236 inch)
Exhaust	5.960 to 5.975 mm (0.234 to 0.235 inch)
Minimum	
Intake	5.945 mm (0.234 inch)
Exhaust	5.930 mm (0.233 inch)
Valve stem clearance	
Standard	
Intake	0.010 to 0.037 mm (0.0004 to 0.0015 inch)
Exhaust	0.025 to 0.052 mm (0.0010 to 0.0020 inch)
Maximum	0.10 mm (0.004 inch)

Valve seat angle 45-degrees
Valve head thickness (intake and exhaust) .... 0.8 to 1.2 mm (0.032 to 0.047 inch)
Valve guide inside diameter (intake and exhaust)
    Standard .... 6.000 to 6.012 mm (0.236 to 0.237 inch)
    Maximum
        Intake .... 6.045 mm (0.238 inch)
        Exhaust .... 6.020 mm (0.237 inch)
Valve seat width (intake and exhaust) .... 0.9 to 1.1 mm (0.035 to 0.043 inch)
Valve face width (intake and exhaust) .... 2.26 mm (0.089 inch)
Valve seat angles
    Top .... 0°
    Seat angle .... 45°
    Bottom .... 60°
Valve spring free length (intake and exhaust)
    Outer spring .... 35.5 mm (1.40 inches)
    Inner spring .... 35.5 mm (1.398 inches)
Valve spring installed height (intake and exhaust) and pressure
    Outer spring (18.5 kg/40.8 lbs) .... 32.0 mm (1.260 inches)
    Inner spring (9.3 kg/20.5 lbs) .... 30.5 mm (1.200 inches)
Valve spring bend limit .... 1.5 mm (0.059 inch)
Valve clearances .... See Chapter 1

Cylinder block

Bore diameter
    Standard .... 58.5 mm (2.303 inches)
    Maximum .... 58.6 mm (2.307 inches)
    Taper and out-of-round limit .... 0.05 mm (0.002 inch)
Bore measuring points .... Top, center and bottom

Pistons

Piston diameter
    Standard .... 57.47 to 58.51 mm (2.302 to 2.304 inches)
    First oversize .... 59.00 mm (2.323 inches)
    Second oversize .... 60.00 mm (2.362 inches)
Diameter measuring point .... 7.0 mm (0.276 inch) from bottom of skirt
Piston-to-cylinder clearance
    Standard .... 0.025 to 0.045 mm (0.0010 to 0.0018 inch)
    Maximum .... 0.15 mm (0.006 inch)
Ring side clearance
    Top ring
        Standard .... 0.03 to 0.07 mm (0.0012 to 0.0028 inch)
        Maximum .... 0.15 mm (0.006 inch)
    Second ring
        Standard .... 0.02 to 0.06 mm (0.0008 to 0.0024 inch)
        Maximum .... 0.15 mm (0.006 inch)
    Oil ring .... Not specified
Ring thickness
    Top ring .... 1.0 mm (0.0394 inch)
    Second ring .... 1.2 mm (0.047 inch)
    Oil ring (spacer and rails) .... 2.5 mm (0.100 inch)
Ring end gap
    Top and second rings
        Standard .... 0.15 to 0.30 mm (0.006 to 0.012 inch)
        Maximum .... 0.7 mm (0.028 inch)
    Oil ring .... 0.20 to 0.70 mm (0.008 to 0.028 inch)

Crankshaft, connecting rods and bearings

Main bearing oil clearance .... 0.021 to 0.044 mm (0.0008 to 0.0017 inch)
Connecting rod side clearance .... 0.160 to 0.262 mm (0.0063 to 0.0103 inch)
Connecting rod bearing oil clearance .... 0.016 to 0.040 mm (0.0006 to 0.0016 inch)
Crankshaft runout limit .... 0.03 mm (0.0012 inch)

Lubrication system

Oil pump
    Inner to outer rotor clearance
        Standard .... 0.09 to 0.15 mm (0.0035 to 0.0059 inch)
        Maximum .... 0.2 mm (0.008 inch)

Lubrication system (continued)

Outer rotor to housing clearance
    Standard .... 0.03 to 0.08 mm (0.0012 to 0.0031 inch)
    Maximum .... 0.15 mm (0.006 inch)
Bypass valve setting pressure .... 0.78 to 1.17 Bars (11.4 to 17 psi
Relief valve opening pressure .... 4.4 to 5.38 Bars (64 to 78.2 psi)
Oil pressure (hot) .... 0.78 Bars (11.4 psi) at 1200 rpm

Clutch

Friction plate thickness
    Standard... 2.9 to 3.1 mm (0.114 to 0.122 inch)
    Minimum.. 2.7 mm (0.106 inch)
Steel plate thickness .... 1.5 to 1.7 mm (0.060 to 0.067 inch)
Steel plate warpage limit .... 0.15 mm (0.006 inch)
Spring free length
    Standard... 42.8 mm (1.69 inches)
    Minimum.. 41.8 mm (1.65 inches)

Transmission

Driveshaft and mainshaft runout limit................................. 0.08 mm (0.0031 inch)
Shift fork guide bar bend limit .... 0.08 mm (0.0031 inch)
Primary reduction ratio .... 22/21 x 65/28 (2.431)
Gear ratios
    1st gear 41/15 (2.733)
    2nd gear 37/19 (1.947)
    3rd gear.. 34/22 (1.545)
    4th gear.. 31/25 (1.240)
    5th gear.. 29/28 (1.036)
    6th gear.. 27/30 (0.900)

Torque specifications

Camshaft sprocket bolts .... 24 Nm (17 ft-lbs)
Cam chain tensioner mounting bolts..................................... 10 Nm (7.2 ft-lbs)
Cam chain tensioner adjusting bolt .... 8 Nm (5.8 ft-lbs)
Adjusting bolt locknut... 9 Nm (6.5 ft-lbs)
Camshaft bearing cap bolts .... 10 Nm (7.2 ft-lbs) (1)
Cylinder block to cylinder head nuts
    M6 diameter .... 10 Nm (7.2 ft-lbs)
    M8 diameter .... 20 Nm (14 ft-lbs)
Cylinder head cap nuts... 22 Nm (16 ft-lbs)
Clutch pressure plate bolts.. 8 Nm (5.8 ft-lbs)
Clutch cover bolts .... 10 Nm (7.2 ft-lbs)
Connecting rod cap nuts .... 25 Nm (18 ft-lbs) (2)
Crankcase plug (M10)... 10 Nm (7.2 ft-lbs)
Crankcase bolts (3)
    6 mm .... 12 Nm (8.7 ft-lbs)
    8 mm .... 24 Nm (17 ft-lbs)
Clutch boss nut .... 70 Nm (50 ft-lbs)
Primary drive gear nut .... 50 Nm (36 ft-lbs)
Starter clutch Allen bolts .... 25 Nm (18 ft-lbs)
Oil pump screws .... 7 Nm (5.1 ft-lbs)
Oil pan bolts .... 10 Nm (7.2 ft-lbs)
Starter chain tensioner bolts.. 10 Nm (7.2 ft-lbs) (4)
Engine mount bolts
    Front upper bolts.. 42 Nm (30 ft-lbs)
    Front bracket bolt .... 32 Nm (23 ft-lbs)
    Front lower bolts .... 42 Nm (30 ft-lbs)
    Rear bolts .... 90 Nm (65 ft-lbs)
    Upper downtube bolts (FZ600) .... 26 Nm (19 ft-lbs)
    Lower downtube bolts (FZ600) .... 40 Nm (29 ft-lbs)
Shift pedal pinch bolt .... 10 Nm (7.2 ft-lbs)
Shift pedal linkage locknuts... 10 Nm (7.2 ft-lbs)
Oil cooler hose fittings .... 32 Nm (23 ft-lbs)
Oil cooler hose bolts.. 12 Nm (8.6 ft-lbs)
Oil cooler mounting bolts .... 10 Nm (7.2 ft-lbs)
Oil cooler hose clamp nuts .... 10 Nm (7.2 ft-lbs)
Oil cooler spacer bolt .... 50 Nm (36 ft-lbs)

Oil pickup bolts..	10 Nm (7.2 ft-lbs)
Shift cam retaining plate screws	10 Nm (7.2 ft-lbs) (4)
Shift cam stopper lever bolt	22 Nm (16 ft-lbs) (4)
Valve cover bolts	10 Nm (7.2 ft-lbs)

1   Tighten evenly in three stages.
2   Apply molybdenum disulphide grease to the threads.
3   Apply engine oil to the threads and tighten in the specified sequence (see text).
4   Apply non-permanent thread locking agent to the threads.

1   General information

The engine/transmission is an air-cooled, in-line four cylinder unit. The valves (one intake and one exhaust per cylinder) are operated by double overhead camshafts which are chain driven off the crankshaft. The engine/transmission assembly is constructed from aluminum alloy. The crankcase is divided horizontally.

The crankcase incorporates a wet sump, pressure-fed lubrication system which uses a gear-driven, single-rotor oil pump, an oil filter, relief valves and an oil level switch. Also contained in the crankcase is the starter motor clutch.

Power from the crankshaft is routed to the transmission via the clutch, which is of the coil spring, wet multi-plate type and is gear-driven off the crankshaft. The transmission is a six-speed, constant-mesh unit.

2   Operations possible with the engine in the frame

The components and assemblies listed below can be removed without having to remove the engine from the frame. If, however, a number of areas require attention at the same time, removal of the engine is recommended.

Gear selector mechanism external components
Starter motor
Alternator
Clutch assembly
Oil pump
Valve cover, camshafts and lifters
Cam chain tensioner
Cylinder head
Cylinder block and pistons

3   Operations requiring engine removal

It is necessary to remove the engine/transmission assembly from the frame and separate the crankcase halves to gain access to the following components:

Oil pan and relief valves
Crankshaft, connecting rods and bearings
Transmission shafts
Shift cam and forks
Camshaft chain and starter chain
Starter clutch and idle gears

4   Major engine repair - general note

1   It is not always easy to determine when or if an engine should be completely overhauled, as a number of factors must be considered.
2   High mileage is not necessarily an indication that an overhaul is needed, while low mileage, on the other hand, does not preclude the need for an overhaul. Frequency of servicing is probably the single most important consideration. An engine that has regular and frequent oil and filter changes, as well as other required maintenance, will most likely give many miles of reliable service. Conversely, a neglected engine, or one which has not been broken in properly, may require an overhaul very early in its life.
3   Exhaust smoke and excessive oil consumption are both indications that piston rings and/or valve guides are in need of attention. Make sure oil leaks are not responsible before deciding that the rings and guides are bad. Refer to Chapter 1 and perform a cylinder compression check to determine for certain the nature and extent of the work required.
4   If the engine is making obvious knocking or rumbling noises, the connecting rod and/or main bearings are probably at fault.
5   Loss of power, rough running, excessive valve train noise and high fuel consumption rates may also point to the need for an overhaul, especially if they are all present at the same time. If a complete tune-up does not remedy the situation, major mechanical work is the only solution.
6   An engine overhaul generally involves restoring the internal parts to the specifications of a new engine. During an overhaul the piston rings are replaced and the cylinder walls are bored and/or honed. If a rebore is done, then new pistons are also required. The main and connecting rod bearings are generally replaced with new ones and, if necessary, the crankshaft is also replaced. Generally the valves are serviced as well, since they are usually in less than perfect condition at this point. While the engine is being overhauled, other components such as the carburetors and the starter motor can be rebuilt also. The end result should be a like-new engine that will give as many trouble free miles as the original.
7   Before beginning the engine overhaul, read through all of the related procedures to familiarize yourself with the scope and requirements of the job. Overhauling an engine is not all that difficult, but it is time consuming. Plan on the motorcycle being tied up for a minimum of two weeks. Check on the availability of parts and make sure that any necessary special tools, equipment and supplies are obtained in advance.
8   Most work can be done with typical shop hand tools, although a number of precision measuring tools are required for inspecting parts to determine if they must be replaced. Often a dealer service department or motorcycle repair shop will handle the inspection of parts and offer advice concerning reconditioning and replacement. As a general rule, time is the primary cost of an overhaul so it doesn't pay to install worn or substandard parts.
9   Any machine shop type operations (valve job, resurfacing, cylinder boring, etc.) should be performed by a specialty shop familiar with motorcycle applications.
10   As a final note, to ensure maximum life and minimum trouble from a rebuilt engine, everything must be assembled with care in a spotlessly clean environment.

5   Engine - removal and installation

Note: *Engine removal and installation should be done with the aid of an assistant to avoid damage or injury that could occur if the engine is dropped. A hydraulic floor jack should be used to support and lower the engine if possible (they can be rented at low cost).*

Removal

Refer to illustrations 5.8, 5.12, 5.14, 5.16a, 5.16b, 5.16c and 5.17

1   Set the bike on its centerstand (if equipped) or support it securely upright.

5.8 Remove the crankcase ventilation hose

5.12 Unbolt the carbon canister and tie it out of the way
(California models only)

5.14 Remove the shift pedal and engine sprocket cover

5.16a Engine mount bolt locations (arrows)

2   Remove the side covers. If you're working on an FJ600 or XJ600, remove the lower fairing. If you're working on an FZ600, remove the lower and center fairing panels (see Chapter 7).
3   Remove the fuel tank (see Chapter 3).
4   Drain the engine oil and remove the oil filter (see Chapter 1).
5   Disconnect both battery cables from the battery. **Warning**: *Always disconnect the negative cable first and reconnect it last to prevent a battery explosion.* Remove the battery and battery case (see Chapter 8).
6   Remove the exhaust system (see Chapter 3).
7   Remove the carburetors (see Chapter 3) and plug the intake openings with rags.
8   Remove the crankcase ventilation hose **(see illustration)**.
9   Refer to Chapters 4 and 8 and disconnect the following electrical connectors:

  a) *Alternator*
  b) *Pickup coil(s)*
  c) *Neutral switch*
  d) *Oil level switch*
  e) *Starter motor*
  f) *Sidestand switch*

10   Remove the starter motor (see Chapter 8). If you're working on a YX600 Radian, remove the horn.
11   Remove the oil cooler and lines, if applicable (see Section 18).

12   If you're working on a California model, unbolt the evaporative emission canister and secure it with a piece of wire so it's out of the way **(see illustration)**.
13   Remove the clutch cable (see Section 20).
14   Remove the shift pedal **(see illustration)**. If you're working on a YX600 Radian, remove the right footpeg bracket and the brake pedal.
15   Remove the engine sprocket cover and engine sprocket (see Chapter 5). It isn't necessary to remove the drive chain completely, but if the engine sprocket is difficult to remove, loosen the rear axle nut and chain adjusters, then push the rear wheel forward to create slack in the chain (see Chapter 1 for details).
16   Remove the engine front and rear mounting bolts and spacer **(see illustrations)**. If you're working on an FZ600, remove the frame downtubes.
17   Place a floor jack and a wood block beneath the oil pan and raise the jack just enough to support the oil pan **(see illustration)**.
18   Make sure no wires or hoses are still attached to the engine assembly. **Warning**: *The engine is heavy and may cause injury if it falls. Be sure it's securely supported. Have an assistant help you steady the engine as you remove it.*

FJ/XJ600 and YX600 Radian models

19   Raise the engine so that the oil pan is level with the frame, then slide the engine out through the right side of the frame and set it onto the floor jack.

5.16b Pull the front bolts out once the nuts are removed . . .

5.16c . . . and then the rear bolt

5.17 Support the engine with a jack and a block of wood

FZ600 models

20   Lower the engine clear of the frame and remove it from the right side of the bike.

All models

21   Slowly and carefully lower the engine assembly to the floor, then guide it out from under the bike.

Installation

22   Installation is the reverse of removal. Note the following points:

a) Apply multipurpose grease to the dampers. don't tighten any of the engine mounting bolts until they all have been installed.
b) Use new gaskets at all exhaust pipe connections.
c) Tighten the engine mount bolts and the downtube-to-frame bolts (on models so equipped) to the torques listed in this Chapter's Specifications.
d) Adjust the drive chain, clutch and throttle cables following the procedures in Chapter 1 and Chapter 3.
e) Refill the engine with oil.

---

6   Engine disassembly and reassembly - general information

Refer to illustrations 6.2a, 6.2b and 6.3

1   Before disassembling the engine, clean the exterior with a degreaser and rinse it with water. A clean engine will make the job

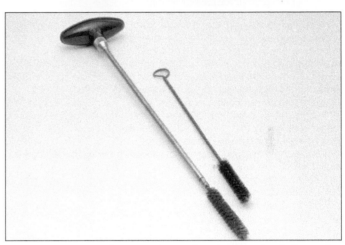

6.2a A selection of brushes is required for cleaning holes and passages in the engine components

6.2b Type HPG-1 Plastigage is needed to check the crankshaft, connecting rod and camshaft oil clearances

easier and prevent the possibility of getting dirt into the internal areas of the engine.

2   In addition to the precision measuring tools mentioned earlier, you will need a torque wrench, a valve spring compressor, valve spring compressor adapter, oil gallery brushes, a piston ring removal and installation tool, a piston ring compressor and a clutch holder tool (which is described in Section 19). Some new, clean engine oil of the correct grade and type, some engine assembly lube (or moly-based grease), a tube of Yamabond (part no. 11001-05-01) or equivalent, and a tube of RTV (silicone) sealant will also be required. Although it may not be considered a tool, some Plastigage (type HPG-1) should also be obtained to use for checking bearing oil clearances (see illustrations).

6.3 An engine stand can be made from short lengths of 2 x 4 lumber and lag bolts or nails

7.5 Remove the valve cover bolts

3   An engine support stand made from short lengths of 2 x 4's bolted together will facilitate the disassembly and reassembly procedures **(see illustration)**. The perimeter of the mount should be just big enough to accommodate the engine oil pan. If you have an automotive-type engine stand, an adapter plate can be made from a piece of plate, some angle iron and a few nuts and bolts.

4   When disassembling the engine, keep "mated" parts together (including gears, cylinders, pistons, valves, etc. that have been in contact with each other during engine operation). These "mated" parts must be reused or replaced as an assembly.

5   Engine/transmission disassembly should be done in the following general order with reference to the appropriate Sections.

Remove the camshafts
Remove the cylinder head
Remove the cylinder block
Remove the pistons
Remove the clutch
Remove the oil pan
Remove the external shift mechanism
Remove the alternator and starter (see Chapter 8)
Separate the crankcase halves
Remove the crankshaft and connecting rods
Remove the transmission shafts/gears
Remove the shift cam/forks
Remove the starter clutch and idle gears

6   Reassembly is accomplished by reversing the general disassembly sequence.

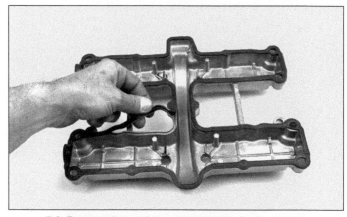

7.9 Be sure the gasket seats securely in the groove

Installation

Refer to illustrations 7.9 and 7.11a through 7.11d

7   Peel the rubber gasket from the cover. If it's cracked, hardened, has soft spots or shows signs of general deterioration, replace it with a new one.

8   Clean the mating surfaces of the cylinder head and the valve cover with lacquer thinner, acetone or brake system cleaner.

9   Install the gasket to the cover. Make sure it fits completely into the cover groove **(see illustration)**.

10   Position the cover on the cylinder head, making sure the gasket doesn't slip out of place.

11   Check the rubber seals on the valve cover bolts, replacing them if necessary **(see illustrations)**. Install the bolts with their seals and washers, tightening them evenly to the torque listed in this chapter's Specifications.

12   The remainder of installation is the reverse of removal.

7   Valve cover - removal and installation

Note: *The valve cover can be removed with the engine in the frame. If the engine has been removed, ignore the steps which don't apply.*

Removal

Refer to illustration 7.5

1   Set the bike on its centerstand (if equipped) or support it securely upright.

2   Remove the seat and side covers (see Chapter 7). If you're working on an FZ600, remove the center and lower fairing panels and the air ducts.

3   Remove the fuel tank (see Chapter 3).

4   Disconnect the spark plug wires from the plugs (see Chapter 1).

5   Remove the valve cover bolts **(see illustration)**.

6   Lift the cover off the cylinder head. If it's stuck, don't attempt to pry it off - tap around the sides with a plastic hammer to dislodge it. **Note**: *Pay attention to the locating dowels as you remove the cover - if they fall into the engine, major disassembly may be required to get them out.*

8   Camshaft chain tensioner - removal and installation

Removal

Refer to illustrations 8.2a, 8.2b and 8.3

1   Loosen the tensioner adjusting bolt locknut while the tensioner is still installed.

2   Remove the tensioner mounting bolts and take it off the engine **(see illustrations)**.

7.11a Inspect the rubber seals on the valve cover bolts and replace them if they're cracked or brittle

7.11b Place the seal and bolt into a socket . . .

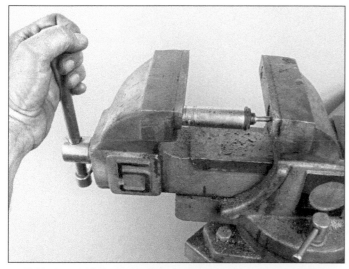

7.11c . . . and then clamp them in a vise to remove the seal

7.11d The new seal can be installed using the same set-up

8.2a Remove the tensioner mounting bolts . . .

8.2b . . . and take the tensioner off

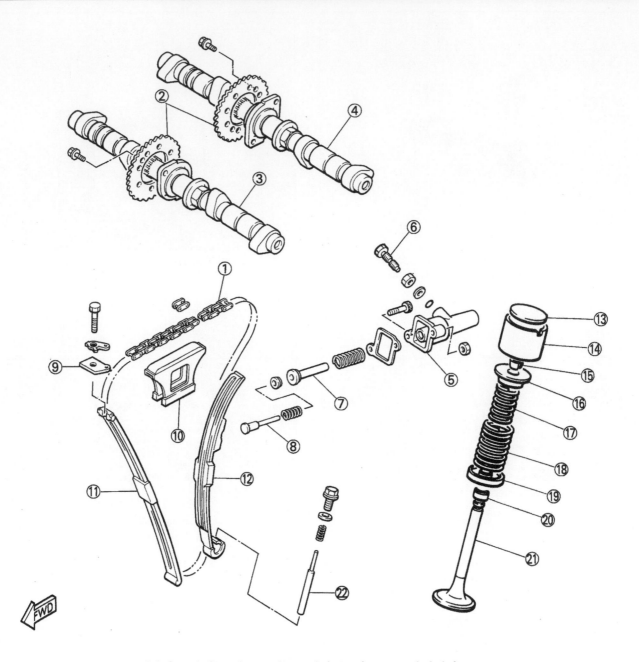

8.3 Camshafts, valves and cam chain tensioner - exploded view

1	Cam chain	9	Chain guide stopper plate	16	Valve spring retainer
2	Cam sprockets	10	Upper chain guide	17	Inner spring
3	Exhaust camshaft	11	Exhaust side chain guide	18	Outer spring
4	Intake camshaft	12	Intake side chain guide	19	Spring seat
5	Chain tensioner body	13	Valve adjusting shim	20	Stem oil seal
6	Tensioner adjusting bolt	14	Valve lifter	21	Valve
7	Large tensioner rod	15	Valve keepers (collets)	22	Intake side chain guide stopper
8	Small tensioner rod				

3   Disassemble the tensioner **(see illustration)**. Check the tensioner parts for wear and damage and replace them as necessary.

Installation
Refer to illustration 8.5

4   Check the sealing washer on the adjusting bolt for cracks or hardening. It's a good idea to replace this washer whenever the tensioner cap is removed.

5   Press the chain tensioner rods into the tensioner housing and lightly tighten the adjusting bolt to hold them there **(see illustration)**.

6   Install the tensioner on the cylinder block, using a new gasket.

7   Tighten the mounting bolts to the torque listed in this Chapter's Specifications.

8   Loosen the tensioner adjusting bolt so the rods can extend.

9   Refer to Chapter 1 and adjust the cam chain tension.

8.5 Push the tensioner rod to the bottom of the housing and snug the adjusting bolt until the tensioner is installed

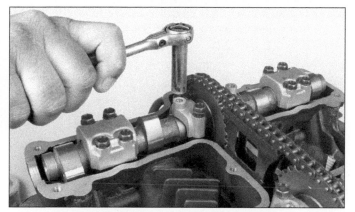

9.5a Remove the no. 3 cam bearing caps. . .

9.5c The camshaft bearing caps are numbered from left to right sides of the engine; with no. 1 cylinder at TDC compression, the no. 1 cam lobes point directly away from each other and the no. 4 cam lobes point directly at each other

1 No. 1 cam bearing caps
2 No. 2 cam bearing caps
3 No. 3 cam bearing cap locations (caps removed)
4 No. 4 cam bearing caps
5 No. 1 cam lobes (no. 1 cylinder at TDC compression)
6 No. 4 cam lobes (no. 1 cylinder at TDC compression)

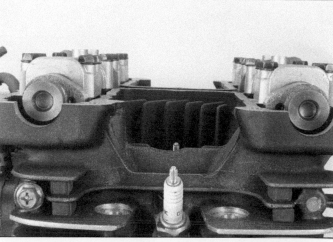

9.3 When the no. 1 cylinder is at TDC on the compression stroke, the cam lobes of no. 4 cylinder should be pointing directly at each other as shown, and the cam lobes of no. 1 cylinder should be pointing directly away from each other

9.5b . . . then remove the cam sprocket bolts and slip the sprockets off the camshafts

---

9 Camshafts and lifters - removal, inspection and installation

Note: *This procedure can be performed with the engine in the frame.*

Camshafts

Removal

Refer to illustrations 9.3, 9.5a, 9.5b, 9.5c, 9.6a, 9.6b, 9.7 and 9.10

1   Set the bike on its centerstand (if equipped) or support it securely upright.

2   Remove the valve cover (see Section 7).

3   Turn the engine to position no. 1 cylinder at TDC compression (see Chapter 1 - *Valve clearances - check and adjustment*). To ease reassembly, make alignment marks on the sprockets, chain and camshafts with a felt pen. The no. 4 cam lobes will face each other when no. 1 cylinder is at TDC **(see illustration)**.

4   Remove the camshaft chain tensioner (see Section 8).

5   Loosen the cap bolts for the no. 3 camshaft bearing caps evenly and remove the no. 3 bearing caps **(see illustration)**. **Note:** *Don't drop the cap bolts or dowels into the engine or major disassembly may be required to get them out. Hold the camshafts from turning with an open end wrench on the camshaft hex and remove the sprocket bolts* **(see illustrations)**. Dismount the sprockets from the camshafts.

9.6a Bend back the lockwasher tab and remove the bolt . . .

9.6b . . . then lift out the exhaust side cam chain guide

9.7 Remove the upper chain guide

9.10 Slip the sprockets off the camshafts, then take the camshafts out

6   Remove the chain guide stopper and lift out the exhaust side chain guide **(see illustrations)**.

7   Pinch the side of the upper cam chain guide and remove from the cylinder head **(see illustration)**.

8   Loosen the camshaft bearing cap bolts for the intake and exhaust camshafts, a little at a time, until all of the bolts are loose. **Caution:** *If the bearing cap bolts aren't loosened evenly, the camshaft may bind. Note that each bearing cap is labeled with a number and the letter I for intake or E for exhaust.*

9   Remove the bolts and lift off the remaining bearing caps and their dowel pins.

10   Slip the camshafts out of the chain, then remove them **(see illustration)**.

11   Remove the camshaft sprockets. Keep the chain taut. Label the sprockets so they can be returned to their original camshafts.

12   While the camshafts are out, don't allow the chain to go slack - the chain may fall and bind between the crankshaft and case, which could damage these components. Install a piece of wire onto the chain to prevent it from dropping down into the crankcase. Also, cover the top of the cylinder head with a rag to prevent foreign objects from falling into the engine.

Inspection

Refer to illustrations 9.13, 9.14a, 9.14b, 9.14c, 9.19a and 9.19b

Note: *Before replacing camshafts or the cylinder head and bearing caps because of damage, check with local machine shops specializing in motorcycle engine work. In the case of the camshafts, it may be possible for cam lobes to be welded, reground and hardened, at a cost far lower than that of a new camshaft. If the bearing surfaces in the cylinder head are damaged, it may be possible for them to be bored out to accept bearing inserts. Due to the cost of a new cylinder head it is recommended that all options be explored before condemning it as trash!*

13   Inspect the cam bearing surfaces of the head and the bearing caps. Look for score marks, deep scratches and evidence of spalling (a pitted appearance) **see illustrations)**.

14   Check the camshaft bearing surfaces and lobes for heat discoloration (blue appearance), score marks, chipped areas, flat spots and spalling **(see illustrations)**. Measure the height of each lobe with a micrometer **(see illustration)** and compare the results to the minimum lobe height listed in this chapter's Specifications. If damage is noted or wear is excessive, the camshaft must be replaced.

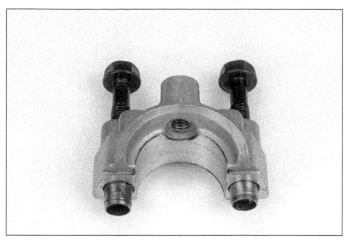

9.13 Inspect the camshaft bearing surfaces in the caps and cylinder head for scratches or wear

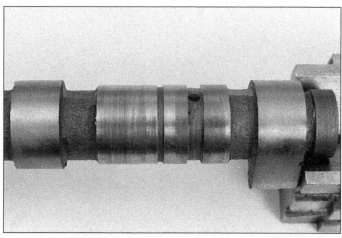

9.14a Check the journal surfaces of the camshaft for scratches or wear

9.14b Check the lobes of the camshaft for wear - here's a good example of damage which will require replacement (or repair) of the camshaft

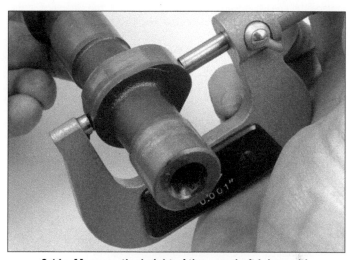

9.14c Measure the height of the camshaft lobes with a micrometer

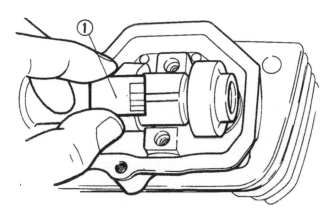

9.19a Compare the width of the crushed Plastigage to the scale on the Plastigage container to obtain the clearance

15   Next, check the camshaft bearing oil clearances. Clean the camshafts, the bearing surfaces in the cylinder head and the bearing caps with a clean, lint-free cloth, then lay the cams in place in the cylinder head.

16   Cut eight strips of Plastigage (type HPG-1) and lay one piece on each bearing journal, parallel with the camshaft centerline.

17   Make sure the bearing cap dowels are installed **(see illustration 9.13)**. Install the bearing caps in their proper positions. The arrows on the caps must face toward the right side of the engine. The numbers on the cap must correspond with the cylinder number (1 through 4, counting from the signal generator end of the engine). The caps labeled I must go on the intake side of the engine and the caps labeled E must go on the exhaust side. Tighten the bolts in three steps to the torque listed in this chapter's Specifications. **Caution:** *Tighten the bearing caps evenly to specifications. While tightening, DO NOT let the camshafts rotate! Use an open end wrench on the hex to hold them steady.*

18   Now unscrew the bolts evenly, a little at a time, and carefully lift off the bearing caps.

19   To determine the oil clearance, compare the crushed Plastigage (at its widest point) on each journal to the scale printed on the Plastigage container **(see illustration)**. Compare the results to this chapter's Specifications. If the oil clearance is greater than specified, measure the diameter of the cam bearing journal with a micrometer

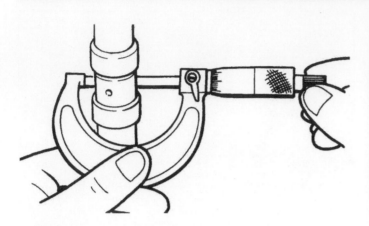

9.19b Measure the camshaft bearing journals with a micrometer

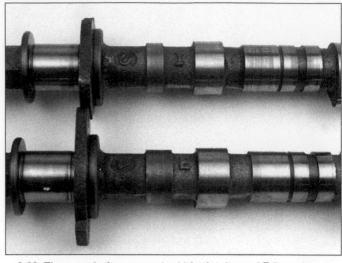

9.26 The camshafts are marked I for intake and E for exhaust

(see illustration). If the journal diameter is less than the specified limit, replace the camshaft with a new one and recheck the clearance. If the clearance is still too great, replace the cylinder head and bearing caps with new parts (see the Note that precedes Step 14).

20   Except in cases of oil starvation, the camshaft chain wears very little. If the chain has stretched excessively, which makes it difficult to maintain proper tension, replace it with a new one (see Section 8).

21   Check the sprockets for wear, cracks and other damage, replacing them if necessary. If the sprockets are worn, the chain is also worn, and also the sprocket on the crankshaft (which can only be remedied by replacing the crankshaft). If wear this severe is apparent, the entire engine should be disassembled for inspection.

22   Check the upper chain guide and exhaust side chain guide for wear or damage. If they're worn or damaged, the chain may be worn out or improperly adjusted. Replacement of the chain requires removal of the crankshaft (see Section 25).

Installation

Refer to illustration 9.26

23   Make sure the crankshaft is still at no. 1 TDC (refer to Chapter 1 - *Valve clearances - check and adjustment*).

24   Slip the camshaft sprockets onto the camshafts but don't install the bolts yet.

25   Make sure the bearing surfaces in the cylinder head and the bearing caps are clean, then apply a light coat of engine assembly lube or moly-based grease to each of them.

26   Apply a coat of moly-based grease to the camshaft lobes. Make sure the camshaft bearing journals are clean, then lay the camshafts in the cylinder head. The camshaft with the I mark goes on the intake side (the same side as the carburetors) and the camshaft with the EX mark goes on the exhaust side **(see illustration)**. Make sure the small punch mark on each camshaft (next to the no. 2 bearing cap) is straight up.

27   Carefully set all of the bearing caps except the no. 3 caps in place with the small arrowhead cast in the top of each cap pointing toward the right (clutch) end of the engine. Make sure the caps are in their proper positions.

 a)   *All caps are marked with the letter I (intake) or E (exhaust). Intake caps go on the intake side of the engine (closest to the carburetors). Exhaust caps go on the exhaust side of the engine (closest to the exhaust ports).*

 b)   *The no. 1 caps (at the left end of the engine) are marked with the number 1.*

 c)   *The no. 2 and no. 3 caps are don't have number marks, but can be distinguished from each other by their shapes.*

 d)   *The no. 4 caps (at the right end of the engine) are marked with the number 4.*

28   Tighten the caps evenly, in three stages, to the torque listed in this Chapter's Specifications.

29   Recheck the camshaft alignment mark (small punch mark) on the exhaust camshaft. It should be aligned with the arrow on top of the no. 2 bearing cap. Once you've done this, engage the sprocket with the cam chain. The sprocket should against its mounting flange on the camshaft, but don't install the sprocket bolts yet.

30   Turn the exhaust sprocket clockwise (viewed from the left end of the engine) to eliminate all slack in the cam chain. While the punch mark on the intake camshaft is aligned with the arrow mark on the no. 2 bearing cap, install the sprocket bolt in the exposed hole and tighten it slightly with a wrench. **Note:** *If the sprocket bolt holes don't line up, reposition the sprocket in the cam chain.*

31   Rotate the intake sprocket clockwise (viewed from the left side of the engine) to take up all the slack in the chain. Make sure the camshaft punch marks are still aligned with the arrow marks on top of the no. 2 bearing caps, then install the sprocket bolt in the exposed hole and tighten it slightly with a wrench. If the bolt holes aren't lined up properly, disengage the sprocket from the chain and make the needed adjustment.

32   recheck to make sure all timing marks - those on the signal generator or alternator rotor, camshafts and no. 2 bearing caps - are aligned correctly. **Caution:** *If the marks are not aligned exactly as described, the valve timing will be incorrect and the valves may contact the pistons, causing extensive damage to the engine. Be sure to recheck the timing mark on the signal generator (early models) or alternator rotor (later models) to make sure it hasn't shifted.*

33   Pour clean engine oil over the cam chain and along the camshafts. Use enough that it flows down onto the sprockets and the valve area.

34   Install the upper cam chain guide, exhaust side cam chain guide and cam chain tensioner.

35   Turn the engine slowly counterclockwise with a wrench on the crankshaft turning bolt (early models) or a socket on the alternator rotor bolt (later models). If you feel a sudden increase in resistance, stop turning. The valves may be hitting the pistons due to incorrect assembly. Find the problem and fix it before turning the engine any further, or serious damage may occur. Once again, check the alignment of the camshaft punch marks with the arrow marks on the no. 2 bearing caps and the crankshaft timing marks.

36   Continue turning until the remaining camshaft sprocket bolt holes are exposed. Install the remaining two sprocket bolts and tighten all four bolts to the torque listed in this Chapter's Specifications.

37   Install the no. 3 bearing caps and tighten their bolt to the torque listed in this Chapter's Specifications.

38   Adjust the cam chain tension (see Chapter 1).

39   The remainder of installation is the reverse of removal.

9.41 Place the lifters in a well marked box so that you don't mix them up

9.42a Mark the lifters and shims with a felt tip pen to identify their locations . . .

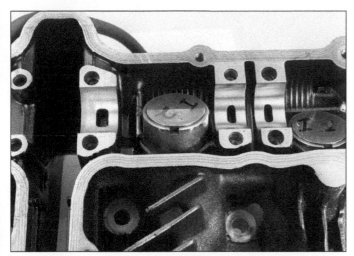

9.42b . . . then remove the lifters from their bores

9.43 Inspect the lifter bores for damage

Valve lifters

Refer to illustrations 9.41, 9.42a and 9.42b

Removal

40   Remove the camshafts following the procedure given above. Be sure to keep tension on the camshaft chain.

41   Make a holder with a separate section for each lifter and its valve adjusting shim (an egg carton or box will work) **(see illustration)**. Label the sections according to cylinder number (1, 2, 3 or 4) and whether the lifter belongs with an intake or exhaust valve. The lifters form a wear pattern with their bores and must be returned to their original locations if reused.

42   Pull the lifter out of the bore with a magnet together with its valve adjusting shim **(see illustrations)**. If the lifters are stuck, spray the area around them with carburetor cleaner and let it soak in. Place the lifters in order in their holder or box.

Inspection

Refer to illustration 9.43

43   Check the lifters and their bores for wear, scuff marks, scratches or other damage **(see illustration)**. Yamaha doesn't provide specifications or wear tolerances for the lifters or their bores, but if the bores are seriously out-of-round or tapered, replace the lifters and cylinder head as a set.

Installation

44   Coat the lifters and their bores with clean engine oil. Place a small dab of moly-based grease or assembly lube onto the adjusting shims prior to camshaft reinstallation.

45   The remainder of installation is the reverse of the removal steps.

10   Cylinder head - removal and installation

Caution: *The engine must be completely cool before beginning this procedure, or the cylinder head may become warped.*

Note: *This procedure can be performed with the engine in the frame. If the engine has been removed, ignore the steps which don't apply.*

Removal

Refer to illustrations 10.8, 10.9, 10.10a, 10.10b, 10.12, 10.13, 10.14a, 10.14b, 10.14c and 10.14d

1   Remove the valve cover (see Section 7).

2   Remove the exhaust system (see Chapter 3).

3   Turn the engine to position no. 1 cylinder at TDC compression (see Chapter 1 - *Valve clearances - check and adjustment*).

4   Remove the cam chain tensioner (see Section 8).

5   Remove the camshafts (see Section 9).

10.8 Remove the nuts and washers from the front of the head . . .

10.9 . . . and from the rear of the head

10.10a Cylinder head nut TIGHTENING sequence

A   Use copper washers

6   Remove the upper cam chain guide (**see illustration 9.7**).
7   Remove the front cam chain guide (**see illustrations 9.5 and 9.6**).
8   Remove the nuts and washers from the front of the cylinder head (**see illustration**).
9   Remove the nuts and washers from the rear of the cylinder head (**see illustration**).
10   Loosen the cylinder head nuts, 1/2 turn at a time, working in the reverse of the tightening sequence (**see illustrations**).
11   Once all of the nuts are loose, remove the nuts and washers. The two washers on the right (clutch) end of the engine are copper. The others are steel.
12   Pull the cylinder head off the cylinder block studs (**see illustration**). If the head is stuck, tap upward with a rubber mallet to jar it loose, or use two wooden dowels inserted into the intake or exhaust ports to rock the head back and forth slightly (its movement will be limited by the studs). Don't attempt to pry the head off by inserting a screwdriver between the head and the cylinder block - you'll damage the sealing surfaces.
13   Lift the head gasket off the cylinder block (**see illustration**). Stuff a clean rag into the cam chain tunnel to prevent the entry of debris.
14   Remove the dowel pins and O-rings (**see illustrations**). don't try to remove the rubber sleeves from the cylinder studs (if equipped) (**see illustration**).
15   Check the cylinder head gasket and the mating surfaces on the cylinder head and block for leakage, which could indicate warpage. Refer to Section 12 and check the flatness of the cylinder head.

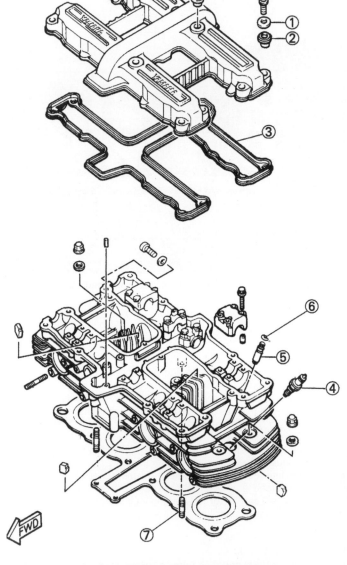

10.10b Cylinder head - exploded view

1 Washer	4 Spark plug	6 Snap-ring
2 Seal	5 Valve guide	7 Stud
3 Gasket		

10.12 Lift the head off the studs

10.13 Lift the head gasket off the cylinder block (it may also have stuck to the head) - always use a new gasket during installation

10.14a Remove the two dowel pins and O-rings from the corners of the cylinder head at the clutch end of the engine

10.14b Install new O-rings during installation

10.14c Tie the timing chain up with wire

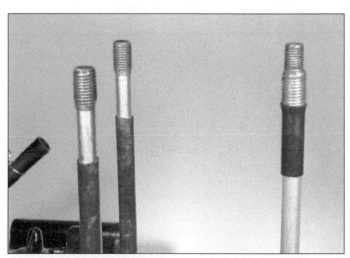

10.14d Don't try to remove the rubber sleeves from the studs (if equipped)

10.18 The HEAD mark on the gasket must be up

12.2a This is one type of adapter used with a standard valve spring compressor to remove the valves

16   Clean all traces of old gasket material from the cylinder head and block. Be careful not to let any of the gasket material fall into the crankcase, the cylinder bores or the oil passages.

Installation

Refer to illustration 10.18

17   Install the dowels on the cylinder head studs. Use new O-rings on the two studs at the end of the head **(see illustration 10.14b)**.

18   Lay the new gasket in place on the cylinder block. If applicable, make sure the HEAD mark on the gasket is up **(see illustration)**. Never reuse the old gasket.

19   Carefully lower the cylinder head over the studs. it's helpful to have an assistant support the camshaft chain with a piece of wire so it doesn't fall and become kinked or detached from the crankshaft. When the head is resting against the cylinder block, wire the cam chain to another component to keep tension on it.

20   Lubricate the threads with engine oil, then install the head washers and nuts. Install new copper washers onto the outer studs on the right side of the engine. Using the proper sequence **(see illustration 10.10a)**, tighten the nuts to approximately half of the torque listed in this chapter's Specifications.

21   Using the same sequence, tighten the nuts to the torque listed in this chapter's Specifications.

22   Lubricate the threads with engine oil, then install the small cylinder block-to-cylinder head nuts, tightening them to the torque listed in this chapter's Specifications.

23   Install the exhaust side and upper cam chain guides.

24   Install the camshafts, tensioner and the valve cover (see Sections 9, 8 and 7).

25   Change the engine oil (see Chapter 1).

26   The remainder of installation is the reverse of the removal steps.

11   Valves/valve seats/valve guides - servicing

1   Because of the complex nature of this job and the special tools and equipment required, servicing of the valves, the valve seats and the valve guides (commonly known as a valve job) is best left to a professional.

2   The home mechanic can, however, remove and disassemble the head, do the initial cleaning and inspection, then reassemble and deliver the head to a dealer service department or properly equipped motorcycle repair shop for the actual valve servicing. Refer to Section 12 for those procedures.

3   The dealer service department will remove the valves and springs, recondition or replace the valves and valve seats, replace the valve guides, check and replace the valve springs, spring retainers and keepers (as necessary), replace the valve seals with new ones and reassemble the valve components.

4   After the valve job has been performed, the head will be in like-new condition. When the head is returned, be sure to clean it again very thoroughly before installation on the engine to remove any metal particles or abrasive grit that may still be present from the valve service operations. Use compressed air, if available, to blow out all the holes and passages.

12   Cylinder head and valves - disassembly, inspection and reassembly

Refer to illustrations 12.2a and 12.2b

1   As mentioned in the previous Section, valve servicing and valve guide replacement should be left to a dealer service department or motorcycle repair shop. However, disassembly, cleaning and inspection of the valves and related components can be done (if the necessary special tools are available) by the home mechanic. This way no expense is incurred if the inspection reveals that service work is not required at this time.

2   To properly disassemble the valve components without the risk of damaging them, a valve spring compressor is absolutely necessary. This special tool can usually be rented, but if it's not available, have a dealer service department or motorcycle repair shop handle the entire process of disassembly, inspection, service or repair (if required) and reassembly of the valves. If your valve spring compressor is not designed to fit inside the lifter bores, you'll need a special adapter to compress the valve springs and retainers in order not to damage the bores **(see illustrations)**.

Disassembly

Refer to illustrations 12.7a and 12.7b

3   Remove the lifters and their shims if you haven't already done so (see Section 9). Store the components so they can be returned to their original locations without getting mixed up.

4   Before the valves are removed, scrape away any traces of gasket material from the head gasket sealing surface. Work slowly and do not nick or gouge the soft aluminum of the head. Gasket removing solvents, which work very well, are available at most motorcycle shops and auto parts stores.

5   Carefully scrape all carbon deposits out of the combustion chamber area. A hand held wire brush or a piece of fine emery cloth can be used once most of the carbon has been scraped away. Do not use a wire brush mounted in a drill motor, or one with extremely stiff

12.2b Attach it to the compressor like this

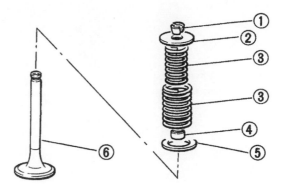

12.7a Valve components - exploded view

1	Valve keepers	4	Stem oil seal
2	Spring retainer	5	Spring seat
3	Valve springs	6	Valve

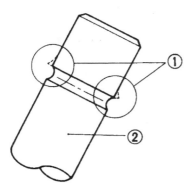

12.7b . . . check the area around the keeper groove for burrs and remove any that you find

1	Burrs (remove with a file)	2	Valve stem

12.14 Lay a precision straightedge across the cylinder head and try to slide a feeler gauge of the specified thickness (equal to the maximum allowable warpage) under it

bristles, as the head material is soft and may be eroded away or scratched by the wire brush. Be careful not to disturb the head surface finish, any scratches or gouges will require that the head be resurfaced.

6   Before proceeding, arrange to label and store the valves along with their related components so they can be kept separate and reinstalled in the same valve guides they are removed from (labeled plastic bags work well for this).

7   Compress the valve spring on the first valve with a spring compressor, then remove the keepers and the retainer from the valve assembly **(see illustration 12.2b and the accompanying illustration)**. Don't compress the springs any more than is absolutely necessary. Carefully release the valve spring compressor and remove the springs and the valve from the head. If the valve binds in the guide (won't pull through), push it back into the head and deburr the area around the keeper groove with a very fine file or whetstone **(see illustration)**.

8   Repeat the procedure for the remaining valves. Remember to keep the parts for each valve together so they can be reinstalled in the same location.

9   Once the valves have been removed and labeled, pull off the valve stem seals with pliers and discard them (the old seals should never be reused), then remove the steel valve spring seats.

10   Next, clean the cylinder head with solvent and dry it thoroughly. Compressed air will speed the drying process and ensure that all holes and recessed areas are clean.

11   Clean all of the valve springs, keepers, retainers and spring seats

with solvent and dry them thoroughly. Do the parts from one valve at a time so that no mixing of parts between valves occurs.

12   Scrape off any deposits that may have formed on the valve, then use a motorized wire brush to remove deposits from the valve heads. Again, make sure the valves don't get mixed up.

Inspection

Refer to illustrations 12.14, 12.15a, 12.15.b, 12.16a, 12.16b, 12.17, 12.18a, 12.18b, 12.19a and 12.19b

13   Inspect the head very carefully for cracks and other damage. If cracks are found, a new head will be required. Check the cam bearing surfaces for wear and evidence of seizure. Check the camshafts and lifters for wear as well (see Section 9).

14   Using a precision straightedge and a feeler gauge, check the head gasket mating surface for warpage. Lay the straightedge lengthwise, across the head and diagonally (corner-to-corner), intersecting the head stud holes, and try to slip a feeler gauge under it, on either side of each combustion chamber **(see illustration)**. The gauge should be the same thickness as the cylinder head warp limit listed in this chapter's Specifications. If the feeler gauge can be inserted between the head and the straightedge, the head is warped and must either be machined or, if warpage is excessive, replaced with a new one.

12.15a Measure the valve seat width with a ruler . . .

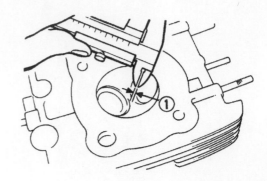

12.15b . . . or for greater precision use a vernier caliper

12.16a Insert a small hole gauge into the valve guide and expand it so there's a slight drag when it's pulled out

12.16b Measure the small hole gauge with a micrometer

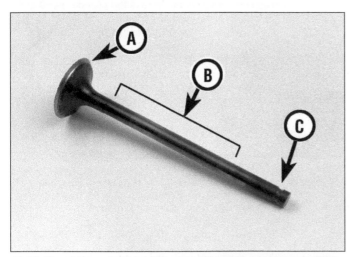

12.17 Check the valve face (A), stem (B) and keeper groove (C) for signs of wear and damage

15   Examine the valve seats in each of the combustion chambers. If they are pitted, cracked or burned, the head will require valve service that's beyond the scope of the home mechanic. Measure the valve seat width **(see illustrations)** and compare it to this chapter's

Specifications. If it is not within the specified range, or if it varies around its circumference, valve service work is required.
16   Clean the valve guides to remove any carbon build-up, then measure the inside diameters of the guides (at both ends and the center of the guide) with a small hole gauge and a 0-to-1-inch micrometer **(see illustrations)**. Record the measurements for future reference. These measurements, along with the valve stem diameter measurements, will enable you to compute the valve stem-to-guide clearance. This clearance, when compared to the Specifications, will be one factor that will determine the extent of the valve service work required. The guides are measured at the ends and at the center to determine if they are worn in a bell-mouth pattern (more wear at the ends). If they are, guide replacement is an absolute must.
17   Carefully inspect each valve face for cracks, pits and burned spots. Check the valve stem and the keeper groove area for cracks **(see illustration)**. Rotate the valve and check for any obvious indication that it is bent. Check the end of the stem for pitting and excessive wear and make sure the bevel is the specified width. The presence of any of the above conditions indicates the need for valve servicing.
18   Measure the valve stem diameter **(see illustration)**. By subtracting the stem diameter from the valve guide diameter, the valve stem-to-guide clearance is obtained. If the stem-to-guide clearance is greater than listed in this chapter's Specifications, the guides and valves will have to be replaced with new ones. Also check the valve stem for bending. Set the valve in a V-block with a dial indicator touching the middle of the stem **(see illustration)**. Rotate the valve and

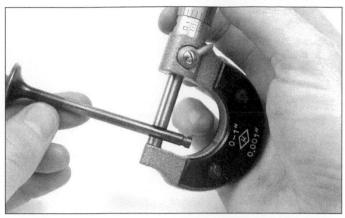

12.18a Measure the valve stem diameter with a micrometer

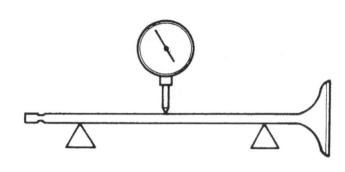

12.18b Check the valve stem for bends with a V-block (or blocks, as shown here) and a dial indicator

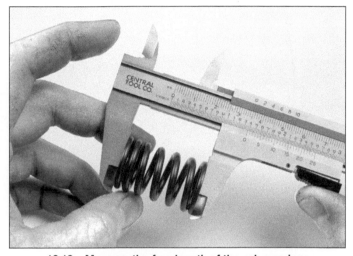

12.19a Measure the free length of the valve springs

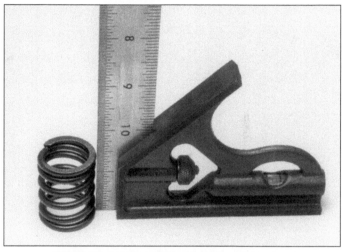

12.19b Check the valve springs for squareness

note the reading on the gauge. If the stem runout exceeds the value listed in this chapter's Specifications, replace the valve.

19   Check the end of each valve spring for wear and pitting. Measure the free length **(see illustration)** and compare it to this chapter's Specifications. Any springs that are shorter than specified have sagged and should not be reused. Stand the spring on a flat surface and check it for squareness **(see illustration)**.

20   Check the spring retainers and keepers for obvious wear and cracks. Any questionable parts should not be reused, as extensive damage will occur in the event of failure during engine operation.

21   If the inspection indicates that no service work is required, the valve components can be reinstalled in the head.

Reassembly

Refer to illustrations 12.23, 12.24a, 12.24b, 12.28a and 12.28b

22   Before installing the valves in the head, they should be lapped to ensure a positive seal between the valves and seats. This procedure requires coarse and fine valve lapping compound (available at auto parts stores) and a valve lapping tool. If a lapping tool is not available, a piece of rubber or plastic hose can be slipped over the valve stem (after the valve has been installed in the guide) and used to turn the valve.

23   Apply a small amount of coarse lapping compound to the valve face **(see illustration)**, then slip the valve into the guide. **Note**: *Make sure the valve is installed in the correct guide and be careful not to get any lapping compound on the valve stem.*

24   Attach the lapping tool (or hose) to the valve and rotate the tool

12.23 Apply the lapping compound very sparingly, in small dabs, to the valve face only

between the palms of your hands. Use a back-and-forth motion rather than a circular motion. Lift the valve off the seat and turn it at regular intervals to distribute the lapping compound properly. Continue the lapping procedure until the valve face and seat contact area is of

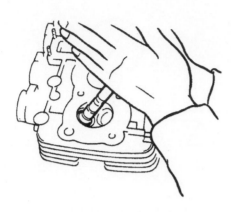

12.24a After lapping, the valve face should exhibit a uniform, unbroken pattern . . .

12.24b . . . and the seat should be the specified width (arrow) with a smooth, unbroken appearance

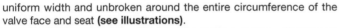

12.28a Install the springs with their closely spaced coils downward (against the cylinder head)

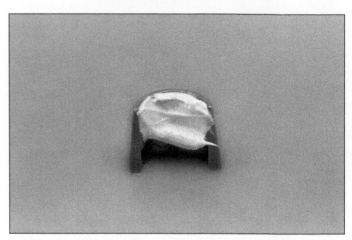

12.28b A small dab of grease will help hold the keepers in place on the valve spring while the valve is released

uniform width and unbroken around the entire circumference of the valve face and seat **(see illustrations)**.

25   Carefully remove the valve from the guide and wipe off all traces of lapping compound. Use solvent to clean the valve and wipe the seat area thoroughly with a solvent soaked cloth.

26   Repeat the procedure with fine valve lapping compound, then repeat the entire procedure for the remaining valves.

27   Lay the spring seats in place in the cylinder head, then install new valve stem seals on each of the guides. Use an appropriate size deep socket to push the seals into place until they are properly seated. don't twist or cock them, or they will not seal properly against the valve stems. Also, don't remove them again or they will be damaged.

28   Coat the valve stems and valve face with assembly lube or moly-based grease, then install one of them into its guide. Next, install the springs and retainers, compress the springs and install the keepers. **Note**: *Install the springs with the tightly wound coils at the bottom (next to the spring seat)* **(see illustration)**. *When compressing the springs with the valve spring compressor, depress them only as far as is absolutely necessary to slip the keepers into place. Apply a small amount of grease to the keepers* **(see illustration)** *to help hold them in place as the pressure is released from the springs. Make certain that the keepers are securely locked in their retaining grooves.*

29   Support the cylinder head on blocks so the valves can't contact the workbench top, then very gently tap each of the valve stems with a soft-faced hammer. This will help seat the keepers in their grooves.

30   Once all of the valves have been installed in the head, check for proper valve sealing by pouring a small amount of solvent into each of the valve ports. If the solvent leaks past the valve(s) into the combustion chamber area, disassemble the valve(s) and repeat the lapping procedure, then reinstall the valve(s) and repeat the check. Repeat the procedure until a satisfactory seal is obtained.

---

13   Cylinder block - removal, inspection and installation

Removal

Refer to illustrations 13.2 and 13.3

1   Following the procedure given in Section 10, remove the cylinder head. Make sure the crankshaft is positioned at Top Dead Center (TDC) for cylinder no. 1.

2   Remove the oil hose clamp from the front of the cylinder block (if equipped). Remove the nut from the stud that secures the block to the crankcase **(see illustration)**.

3   Lift the cylinder block straight up to remove it **(see illustration)**. If it's stuck, tap around its perimeter with a soft-faced hammer. Don't attempt to pry between the block and the crankcase, as you will ruin the sealing surfaces. As you lift, note the location of the dowel pins. Be careful not to let these drop into the engine.

4   Stuff clean shop towels around the pistons and remove the gasket and all traces of old gasket material from the surfaces of the cylinder block and the cylinder head.

13.2 Remove the nut from the single stud at the front of the cylinder block

13.3 Lift the cylinder block up and off the crankcase and pistons

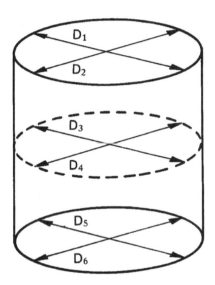

13.7 Measure the cylinder bore with a telescoping gauge at these six points (then measure the gauge with a micrometer) . . .

13.12 Use a scraper to remove any remaining gasket material from the bottom of the cylinder block

Inspection

Refer to illustration 13.7

5   don't attempt to separate the liners from the cylinder block.

6   Check the cylinder walls carefully for scratches and score marks.

7   Using the appropriate precision measuring tools, check each cylinder's diameter. On all models, measure near the top, center and bottom of the cylinder bore, parallel to the crankshaft axis **(see illustration)**. Next, measure each cylinder's diameter at the same three locations across the crankshaft axis. Compare the results to this chapter's Specifications. If the cylinder walls are tapered, out-of-round, worn beyond the specified limits, or badly scuffed or scored, have them rebored and honed by a dealer service department or a motorcycle repair shop. If a rebore is done, oversize pistons and rings will be required as well.

8   As an alternative, if the precision measuring tools are not available, a dealer service department or motorcycle repair shop will make the measurements and offer advice concerning servicing of the cylinders.

9   If they are in reasonably good condition and not worn to the outside of the limits, and if the piston-to-cylinder clearances can be maintained properly (see Section 14), then the cylinders do not have to be rebored; honing is all that is necessary.

10   To perform the honing operation you will need the proper size flexible hone with fine stones, or a bottle-brush type hone, plenty of light oil or honing oil, some shop towels and an electric drill motor. Hold the cylinder block sideways (bore 90° to the vise jaws) in a vise (cushioned with soft jaws or wood blocks) when performing the honing operation. Mount the hone in the drill motor, compress the stones and slip the hone into the cylinder. Lubricate the cylinder thoroughly, turn on the drill and move the hone up and down in the cylinder at a pace which will produce a fine crosshatch pattern on the cylinder wall with the crosshatch lines intersecting at approximately a 60-degree angle. Be sure to use plenty of lubricant and do not take off any more material than is absolutely necessary to produce the desired effect. Do not withdraw the hone from the cylinder while it is running. Instead, shut off the drill and continue moving the hone up and down in the cylinder until it comes to a complete stop, then compress the stones and withdraw the hone. Wipe the oil out of the cylinder and repeat the procedure on the other cylinders. Remember, do not remove too much material from the cylinder wall. If you do not have the tools, or do not desire to perform the honing operation, a dealer service department or motorcycle repair shop will generally do it for a reasonable fee.

11   Next, the cylinders must be thoroughly washed with warm soapy water to remove all traces of the abrasive grit produced during the honing operation. Be sure to run a brush through the bolt holes and flush them with running water. After rinsing, dry the cylinders thoroughly and apply a coat of light, rust-preventative oil to all machined surfaces.

Installation

Refer to illustrations 13.12, 13.13a through 13.13d and 13.16

12   Ensure that the bottom gasket surface is clean **(see illustration)** before you lubricate the cylinder bores with clean engine oil. Apply a thin film of engine oil to the piston skirts.

13.13a Install a new O-ring around the base of each cylinder . . .

13.13b . . . then, install the new gasket between the block and cylinder assembly . . .

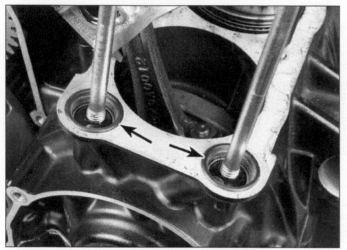

13.13c . . . and new O-rings at the clutch end of the engine

13   Install the dowel pins, then place a new cylinder base gasket on the cylinder block. Install new O-rings on the base of each cylinder. Install new O-rings around the cylinder block dowels **(see illustrations)**.

14   Slowly rotate the crankshaft until two of the pistons are up and two down. Be extremely careful not to jam the timing chain in the case.

15   Attach four piston ring compressors to the pistons and compress the piston rings. Large hose clamps can be used instead - just make sure they don't scratch the pistons, and don't tighten them too much.

16   Install the cylinder block over the pistons and carefully lower it down until the piston crowns fit into the cylinder liners **(see illustration)**. While doing this, pull the camshaft chain up, using a hooked tool or a piece of coat hanger or previously attached piece of wire. Also keep an eye on the cam chain guide to make sure it doesn't wedge against the block. Push down on the cylinder block, making sure the pistons don't get cocked sideways, until the bottoms of the cylinder liners slide down past the piston rings. A wood or plastic hammer handle can be used to gently tap the block down, but don't use too much force or the pistons will be damaged.

17   Once the cylinders have passed over the rings of the two pistons, rotate the crankshaft so the remaining pistons are up and install the cylinder block over them.

18   Remove the piston ring compressors or hose clamps (if used), being careful not to scratch the pistons. Remove the rods from under the pistons.

19   The remainder of installation is the reverse of removal.

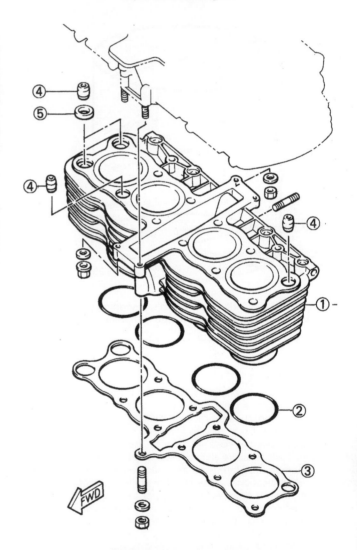

13.13d Cylinder block details

1   Cylinder block   4   Dowels
2   O-ring   5   O-ring
3   Gasket

13.16 If you're experienced and very careful, you can install the cylinder block over the pistons without using ring compressors, but it's advised to use them

14.3a Mark the cylinder numbers on the piston crowns (if the top surface needs cleaning, use a sharp scribe so you don't accidentally remove the number) - also note the arrow, which must point to the front

14.3b Remove the circlip from one side of the piston - wear eye protection and be careful not to let it fly out

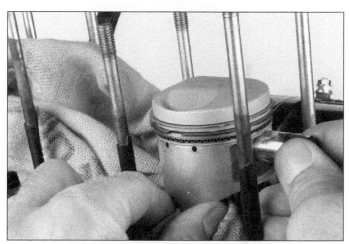

14.4a Push the piston pin out until you can grasp it, then pull it out the rest of the way

14 Pistons - removal, inspection and installation

1   The pistons are attached to the connecting rods with piston pins that are a slip fit in the pistons and rods.
2   Before removing the pistons from the rods, stuff a clean shop towel into each crankcase hole, around the connecting rods. This will prevent the circlips from falling into the crankcase if they are inadvertently dropped.

Removal

Refer to illustrations 14.3a, 14.3b, 14.4a and 14.4b

3   Using a sharp scribe, scratch the number of each piston into its crown (or use a felt pen if the piston is clean enough). Each piston should also have an arrow pointing toward the front of the engine (see illustration). If not, scribe an arrow into the piston crown before removal. Support the first piston, grasp the circlip with needle-nose pliers and remove it from the groove (see illustration).
4   Push the piston pin out from the opposite end to free the piston from the rod (see illustration). You may have to deburr the area around the groove to enable the pin to slide out (use a triangular file for this procedure). If the pin won't come out, remove the remaining circlip. Fabricate a piston pin removal tool from threaded stock, nuts, washers and a piece of pipe (see illustration). Repeat the procedure for the other pistons.

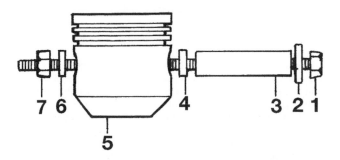

14.4b The piston pins should come out with hand pressure - if they don't, this removal tool can be fabricated from readily available parts

1 Bolt	5 Piston
2 Washer	6 Washer (B)
3 Tubing or pipe (A)	7 Nut (B)
4 Padding (A)	

14.6 Remove the piston rings with a ring removal and installation tool

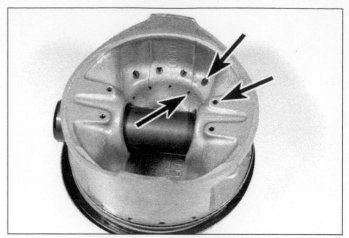

14.11 Check the piston pin bore and the piston skirt for wear, and make sure the internal holes are clear (arrows)

14.13 Measure the piston ring-to-groove clearance with a feeler gauge

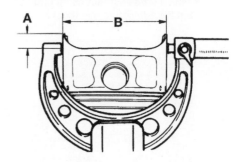

14.14 Measure the piston diameter with a micrometer

A   *Specified distance from bottom of piston*
B   *Piston diameter*

Inspection

Refer to illustrations 14.6, 14.11, 14.13, 14.14 and 14.17

5   Before the inspection process can be carried out, the pistons must be cleaned and the old piston rings removed.

6   Using a piston ring removal and installation tool, carefully remove the rings from the pistons **(see illustration)**. Do not nick or gouge the pistons in the process.

7   Scrape all traces of carbon from the tops of the pistons. A hand-held wire brush can be used once most of the deposits have been scraped away. Do not, under any circumstances, use a wire brush mounted in a drill motor to remove deposits from the pistons; the piston material is soft and will be eroded away by the wire brush.

8   Use a piston ring groove cleaning tool to remove any carbon deposits from the ring grooves. If a tool is not available, a piece broken off the old ring will do the job. Be very careful to remove only the carbon deposits. Do not remove any metal and do not nick or gouge the top and bottom of the ring lands (the metal ridges between the ring grooves).

9   Once the deposits have been removed, clean the pistons with solvent and dry them thoroughly. Make sure the oil return holes below the oil ring grooves are clear.

10   If the pistons are not damaged or worn excessively and if the cylinders are not rebored, new pistons will not be necessary. Normal piston wear appears as even, vertical wear on the thrust surfaces of the piston and slight looseness of the top ring in its groove. New piston rings, on the other hand, should always be used when an engine is rebuilt.

11   Carefully inspect each piston for cracks around the skirt, at the pin bosses and at the ring lands **(see illustration)**.

12   Look for scoring and scuffing on the thrust faces of the skirt, holes in the piston crown and burned areas at the edge of the crown. If the skirt is scored or scuffed, the engine may have been suffering from overheating and/or abnormal combustion, which caused excessively high operating temperatures. The oil pump and oil cooler (if equipped) should be checked thoroughly. A hole in the piston crown, an extreme to be sure, is an indication that abnormal combustion (pre-ignition) was occurring. Burned areas at the edge of the piston crown are usually evidence of spark knock (detonation). If any of the above problems exist, the causes must be corrected or the damage will occur again.

13   Measure the piston ring-to-groove clearance by laying a new piston ring in the ring groove and slipping a feeler gauge in beside it **(see illustration)**. Check the clearance at three or four locations around the groove. Be sure to use the correct ring for each groove; they are different. If the clearance is greater than specified, new pistons will have to be used when the engine is reassembled.

14   Check the piston-to-bore clearance by measuring the bore (see Section 13) and the piston diameter. Make sure that the pistons and cylinders are correctly matched. Measure the piston across the skirt on the thrust faces at a 90-degree angle to the piston pin, at the distance from the bottom of the skirt listed in this Chapter's Specifications **(see illustration)**. Subtract the piston diameter from the bore diameter to obtain the clearance. If it is greater than specified, the cylinders will have to be rebored and new oversized pistons and rings installed.

15   If the appropriate precision measuring tools are not available, the piston-to-cylinder clearances can be obtained, though not quite as accurately, using feeler gauge stock. Feeler gauge stock comes in 12-inch lengths and various thicknesses and is generally available at auto parts stores. To check the clearance, select a feeler gauge of the same thickness as the piston clearance listed in this Chapter's Specifications and slip it into the cylinder along with the appropriate piston. The cylinder should be

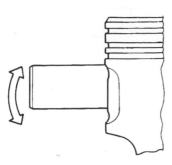

14.17 Slip the pin into the piston and try to rock it back-and-forth; if it's loose, replace the piston and pin

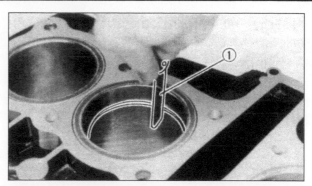

15.3 Square the ring in the bore by turning the piston upside down and tapping on the ring, then check the piston ring end gap with a feeler gauge

15.5 If the end gap is too small, clamp a file in a vise and file the ring ends (from the outside in only) to enlarge the gap slightly

15.9a Installing the oil ring expander - make sure the ends don't overlap

upside down and the piston must be positioned exactly as it normally would be. Place the feeler gauge between the piston and cylinder on one of the thrust faces (90-degrees to the piston pin bore). The piston should slip through the cylinder (with the feeler gauge in place) with moderate pressure. If it falls through, or slides through easily, the clearance is excessive and a new piston will be required. If the piston binds at the lower end of the cylinder and is loose toward the top, the cylinder is tapered, and if tight spots are encountered as the feeler gauge is placed at different points around the cylinder, the cylinder is out-of-round.

16   Repeat the procedure for the remaining pistons and cylinders. Be sure to have the cylinders and pistons checked by a dealer service department or a motorcycle repair shop to confirm your findings before purchasing new parts.

17   Apply clean engine oil to the pin, insert it into the piston and check for freeplay by rocking the pin back-and-forth **(see illustration)**. If the pin is loose, new pistons and pins must be installed.

18   Refer to Section 15 and install the rings on the pistons.

Installation

19   Install the pistons in their original locations with the arrows pointing to the front of the engine. Lubricate the pins and the rod bores with clean engine oil. Install new circlips in the grooves in the inner sides of the pistons (don't reuse the old circlips). Push the pins into position from the opposite side and install new circlips. Compress the circlips only enough for them to fit in the piston. Make sure the clips are properly seated in the grooves and that the open of the circlip is away from the removal notch.

15 Piston rings - installation

Refer to illustrations 15.3, 15.5, 15.9a, 15.9b, 15.11, 15.12 and 15.15

1   Before installing the new piston rings, the ring end gaps must be checked.

2   Lay out the pistons and the new ring sets so the rings will be matched with the same piston and cylinder during the end gap measurement procedure and engine assembly.

3   Insert the top (No. 1) ring into the bottom of the first cylinder and square it up with the cylinder walls by pushing it in with the top of the piston. Since the bottom of the cylinder experiences the least amount of wear, the ring should be about one inch above the bottom edge of the cylinder. To measure the end gap, slip a feeler gauge between the ends of the ring **(see illustration)** and compare the measurement to this Chapter's Specifications.

4   If the gap is larger or smaller than specified, double check to make sure that you have the correct rings before proceeding.

5   If the gap is too small, it must be enlarged or the ring ends may come in contact with each other during engine operation, which can cause serious damage. The end gap can be increased by filing the ring ends very carefully with a fine file **(see illustration)**. When performing this operation, file only from the outside in.

6   Excess end gap is not critical unless it is greater than 0.040 in (1 mm). Again, double check to make sure you have the correct rings for your engine. Too little end gap can cause the rings to seize to the cylinder walls.

7   Repeat the procedure for each ring that will be installed in the first cylinder and for each ring in the remaining cylinders. Remember to keep the rings, pistons and cylinders matched up.

8   Once the ring end gaps have been checked/corrected, the rings can be installed on the pistons.

9   The oil control ring (lowest on the piston) is installed first. It is composed of three separate components. Slip the expander into the groove, then install the upper side rail **(see illustrations)**. Don't use a piston ring installation tool on the oil ring side rails as they may be damaged. Instead, place one end of the side rail into the groove

15.9b Installing an oil ring side rail - don't use a ring installation tool to do this

15.15 Arrange the ring gaps like this

1	Top compression ring	3	Oil ring upper rail
2	Oil ring lower rail	4	Second compression ring

between the spacer expander and the ring land. Hold it firmly in place and slide a finger around the piston while pushing the rail into the groove. Next, install the lower side rail in the same manner.

10   After the three oil ring components have been installed, check to make sure that both the upper and lower side rails can be turned smoothly in the ring groove.

11   Install the second (middle) ring next. It can be readily distinguished from the top ring by its cross-section shape **(see illustration)**. Do not mix the top and middle rings.

12   To avoid breaking the ring, use a piston ring installation tool and make sure that the identification mark is facing up **(see illustration)**. Fit the ring into the middle groove on the piston. Do not expand the ring any more than is necessary to slide it into place.

13   Finally, install the top ring in the same manner. Make sure the identifying mark is facing up.

14   Repeat the procedure for the remaining pistons and rings. Be very careful not to confuse the top and second rings.

15   Once the rings have been properly installed, stagger the end gaps, including those of the oil ring side rails **(see illustration)**.

16   Apply a liberal coating of engine oil onto the pistons prior to installation.

16   Oil pan and relief valves - removal, relief valve inspection and installation

Removal

Refer to illustrations 16.2, 16.4, 16.5a, 16.5b, 16.7a and 16.7b

1   Remove the engine (see Section 5).

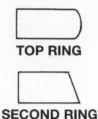

15.11 Don't confuse the top ring with the second ring

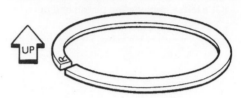

15.12 Make sure the marks on the rings face up when the rings are installed on the pistons

2   Remove the oil pan bolts and detach the pan from the crankcase **(see illustration)**.

3   Remove all traces of old gasket material from the mating surfaces of the oil pan and crankcase.

4   Remove the strainer from the oil pickup and inspect for debris **(see illustration)**.

5   Unbolt the pickup from the crankcase and remove the pickup gasket **(see illustrations)**.

6   Remove all traces of old gasket from the pickup and crankcase.

7   Remove the relief valves from the crankcase **(see illustrations)**. The oil pressure valve should pull out with light hand pressure. If it's stuck, rock it back and forth slightly while you pull on it. The starter chain tensioner relief valve should be unscrewed with a wrench or socket.

Relief valve inspection

Refer to illustrations 16.8, 16.9, 16.10a, 16.10b, 16.10c and 16.12

8   Push the plunger into the relief valve and check for free movement **(see illustration)**. If the valve sticks, perform the following steps to disassemble and inspect it.

16.2 Remove the oil pan bolts around the edge of the pan

16.4 Remove the oil strainer from the pickup; the arrows on strainer and pickup point toward the front of the engine

16.5a Remove the pickup bolts . . .

16.5b . . . and detach the pickup from the engine - replace the pickup gasket with a new one

16.7a Pull the main oil relief valve out of the crankcase . . .

16.7b . . . and unscrew the starter chain relief valve from the block

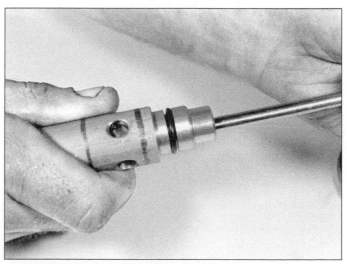

16.8 Push in on the relief valve plunger to make sure it moves freely

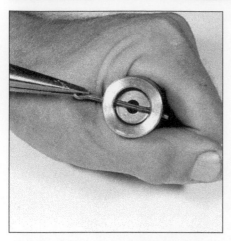

16.9 Straighten the ends of the cotter pin
and pull it out

16.10a Remove the spring retainer . . .

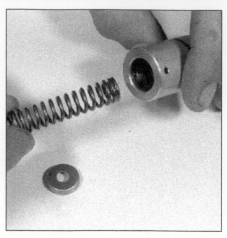

16.10b . . . the spring . . .

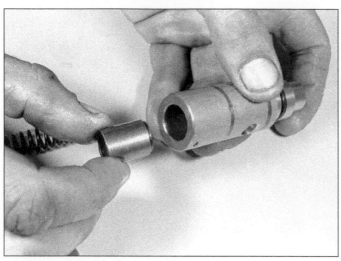

16.10c . . . and the plunger

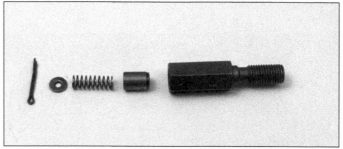

16.12 Starter chain tensioner relief valve details

17 Oil pump - pressure check, removal, inspection and installation

Note: *The oil pump can be removed with the engine in the frame.*

Oil pressure check

Refer to illustration 17.2

Warning: *If the oil passage plug is removed when the engine is hot, hot oil will drain out - wait until the engine is cold before beginning this check.*

1   Remove the lower fairing (if equipped) (see Chapter 7).
2   Remove the plug at the bottom of the crankcase on the left side and install an oil pressure gauge **(see illustration)**.

17.2 To check the oil pressure, remove the plug (arrow) and
connect an oil pressure gauge using the proper adapter

9   Straighten the cotter pin and pull it out **(see illustration)**.
10   Remove the spring retainer, spring and plunger **(see illustrations)**.
11   Check all parts for wear and damage. Clean the parts thoroughly, reassemble the valve and recheck its movement. If the valve still sticks, replace it. **Caution**: *If you reuse the relief valve, install a new cotter pin before reinstalling the relief valve in the engine.*
12   Repeat the same steps for the starter chain tensioner relief valve **(see illustration)**.

Installation

13   Install a new oil pickup gasket **(see illustration 16.5b)**.
14   Install the oil pan dowels. Install the relief valves in the crankcase, using new O-rings.
15   Install the oil pickup and tighten the bolts to the torque listed in this Chapter's Specifications. Install the strainer on the pickup with its open end toward the rear of the engine.
16   Position a new gasket on the oil pan. A thin film of sealant can be used to hold the gasket in place. Install the oil pan and bolts, tightening the bolts to the torque listed in this chapter's Specifications, using a criss-cross pattern. Two of the bolts secure wiring harness clips.
17   The remainder of installation is the reverse of removal. Install a new oil filter and fill the crankcase with oil (see Chapter 1), then run the engine and check for leaks.

3   Start the engine and watch the gauge while varying the engine rpm. The pressure should stay within the relief valve opening pressure listed in this chapter's Specifications. If the pressure is too high, the relief valve(s) is stuck closed. To check it, see Section 16.

4   If the pressure is lower than specified, either a relief valve is stuck open, the oil pump is faulty, or there is other engine damage. Begin diagnosis by checking the relief valves (see Section 16), then the oil pump. If those items check out okay, chances are the bearing oil clearances are excessive and the engine needs to be overhauled.

5   If the pressure reading is in the desired range, allow the engine to warm up to normal operating temperature and check the pressure again, at the specified engine rpm. Compare your findings with this chapter's Specifications.

6   If the pressure is significantly lower than specified, check the relief valves and the oil pump.

Removal

Refer to illustrations 17.8a, 17.8b, 17.9a and 17.9b

7   Remove the clutch (see Section 19).

8   Remove the snap-ring and take off the oil pump driven gear **(see illustrations)**.

9   Remove the oil pump mounting screws and lift the pump away from the engine **(see illustration)**. Remove the O-rings **(see illustration)**.

17.8a **Remove the snap-ring and the oil pump drive gear**

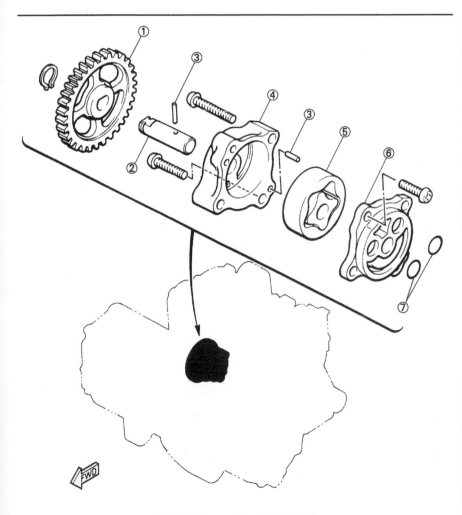

17.8b **Oil pump (exploded view)**

1   *Drive gear*   5   *Inner and outer rotors*
2   *Rotor shaft*   6   *Pump cover*
3   *Pins*   7   *O-rings*
4   *Pump body*

17.9a **Remove the three mounting screws (arrows) and take the pump off**

17.9b **Remove the O-rings (arrows)**

17.12 Remove the cover screw and lift off the cover

17.14 Insert a feeler gauge between the outer rotor and pump body to measure the clearance

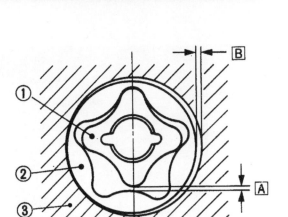

17.15 Measure the clearance between the outer rotor tip and the inner rotor with a feeler gauge

1   Inner rotor
2   Outer rotor
3   Pump body
A   Inner rotor to outer rotor clearance
B   Outer rotor to pump body clearance

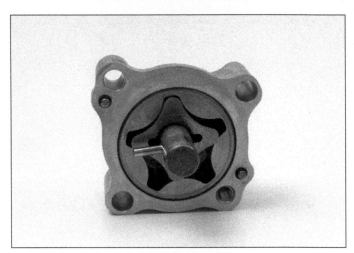

17.16 Insert the pin in the shaft and make sure it's centered

Inspection

Refer to illustrations 17.12, 17.14, 17.15 and 17.16

10   Wash the oil pump in solvent, then dry it off.

11   Remove the pump cover screw (use an impact driver if it's tight) **(see illustration 17.8b)**.

12   Lift off the cover **(see illustration)**.

13   Check the pump body and rotors for scoring and wear. If any damage or uneven or excessive wear is evident, replace the pump (individual parts aren't available). If you're rebuilding the engine, it's a good idea to install a new oil pump.

14   Measure the clearance between the outer rotor and pump body with a feeler gauge and compare it to the value listed in this chapter's Specifications **(see illustration)**. If it's excessive, replace the pump.

15   Measure the clearance between the inner and outer rotors**(see illustration)**. Again, replace the pump if the clearance is excessive.

16   If the pump is good, reverse the disassembly steps to reassemble it. Make sure the pin is centered in the rotor shaft so it will align with the slot in the inner rotor **(see illustration)**.

Installation

17   Before installing the pump, prime it by pouring oil into it while turning the shaft by hand - this will ensure that it begins to pump oil quickly.

18   Installation is the reverse of removal, with the following additions:

a) *Make sure the O-rings are in place.*

b) *Tighten the pump mounting screws to the torque listed in this chapter's Specifications.*

18   Oil cooler and hoses - removal and installation

Refer to illustration 18.3

1   Remove the lower fairing (see Chapter 7) and drain the engine oil (see Chapter 1).

2   Remove the bolts that secure the oil cooler hoses to the oil filter adapter. Should the adapter require service, always install new O-rings upon reassembly.

3   Detach the oil cooler hose retainer from the front of the engine **(see illustrations)**.

4   Remove the oil cooler mounting bolts.

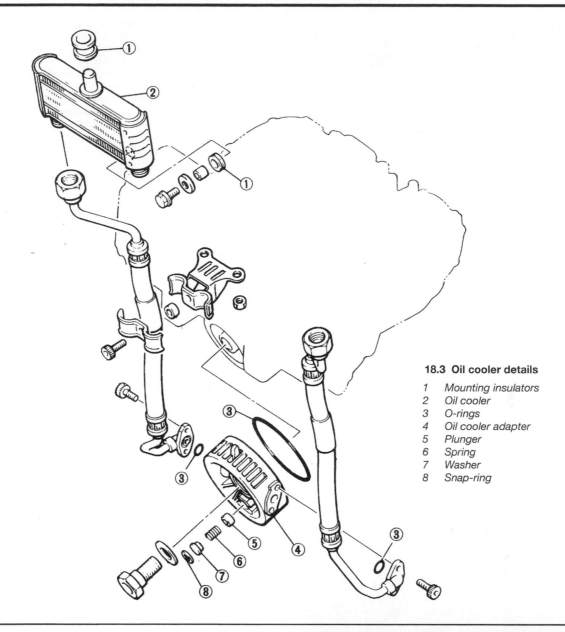

18.3 Oil cooler details

1   *Mounting insulators*
2   *Oil cooler*
3   *O-rings*
4   *Oil cooler adapter*
5   *Plunger*
6   *Spring*
7   *Washer*
8   *Snap-ring*

5   Installation is the reverse of the removal steps, with the following additions:

a) *Use new O-rings between the hoses and the oil filter adapter.*
b) *Tighten the bolts and hose fittings to the torques listed in this chapter's Specifications.*

6   Fill the engine with oil (see Chapter 1).
7   Run the engine and check for leaks.

---

19   Clutch - removal, inspection and installation

Note: *The clutch can be removed with the engine in the frame.*

Removal

Refer to illustrations 19.3, 19.4a, 19.4b, 19.7, 19.9a, 19.9b, 19.9c, 19.10 and 19.11

1   Set the bike on its centerstand (if equipped) or prop it securely upright. Remove the lower fairing (FJ/XJ600) or lower and center fairing panels (FZ600) (see Chapter 7).
2   Drain the engine oil (see Chapter 1).
3   Remove the clutch cover **(see illustration)**.

19.3 Loosen the clutch cover Allen bolts evenly in a criss-cross pattern

19.4a Remove the pressure plate bolts in a criss-cross pattern; note the location of the reference mark on the pressure plate (arrow)

4   Loosen the clutch pressure plate bolts evenly in a criss-cross pattern, then remove the bolts and washers **(see illustrations)**.
5   Remove the clutch springs.
6   Remove the pressure plate.
7   Remove the pull rod, thrust bearing and plate washer from the pressure plate **(see illustration)**.
8   Remove the friction plates and steel plates as a set. Note the position
9   Bend back the lockwasher tab on the clutch boss nut **(see illustration)**. Remove the nut, using a special holding tool or equivalent (Yamaha tool no. YM-91402 in the US; 90890-04086 in the UK) to prevent the clutch housing from turning **(see illustration)**. An alternative to this tool can be fabricated from some steel strap, bent at the ends and bolted together in the middle **(see illustration)**. Remove the lockwasher and discard it. Use a new one during installation.
10   Remove the outer clutch boss and rear thrust washer **(see illustration)**.
11   Remove the snap ring to remove the oil pump drive gear **(see illustration)**.

Inspection

Refer to illustrations 19.12, 19.14, 19.15, 19.19 and 19.21
12   Examine the splines on both the inside and the outside of the

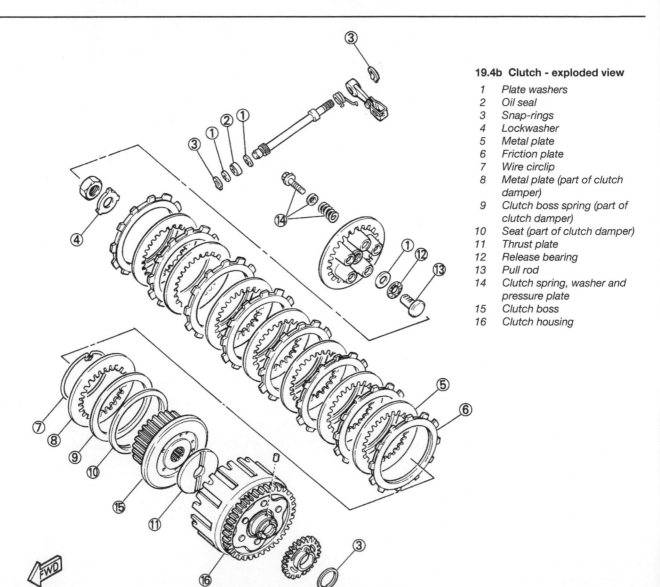

19.4b Clutch - exploded view

1   *Plate washers*
2   *Oil seal*
3   *Snap-rings*
4   *Lockwasher*
5   *Metal plate*
6   *Friction plate*
7   *Wire circlip*
8   *Metal plate (part of clutch damper)*
9   *Clutch boss spring (part of clutch damper)*
10   *Seat (part of clutch damper)*
11   *Thrust plate*
12   *Release bearing*
13   *Pull rod*
14   *Clutch spring, washer and pressure plate*
15   *Clutch boss*
16   *Clutch housing*

19.7 Remove the pull rod, release bearing and washer
from the pressure plate

19.9a Bend the tab on the clutch boss lockwasher away
from the nut, then remove the nut

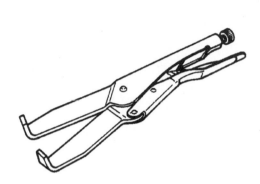

19.9b You'll need a tool to keep the clutch from turning; this is
the Yamaha special tool . . .

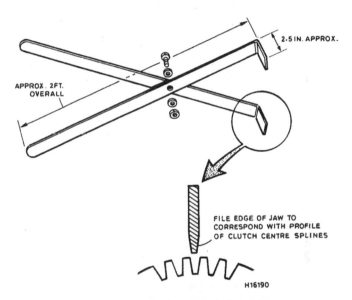

19.9c . . . or you can make your own tool from steel scrap

19.10 Slide the clutch boss off and remove the thrust washer

19.11 Remove the snap-ring (arrow) and take off the
oil pump drive gear

19.12 Check the clutch boss for wear and damage, paying
special attention to the splines

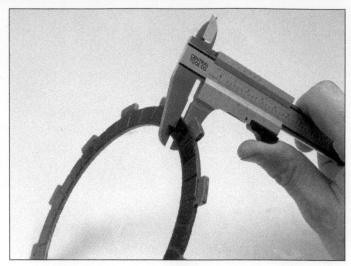

19.14 Measure the thickness of the friction plates

19.15 Check the metal plates for warpage

19.19 Check the release bearing for worn or damaged rollers

clutch hub (see illustration). If any wear is evident, replace the clutch hub with a new one.

13   Measure the free length of the clutch springs. Replace the springs as a set if any are not within the values listed in this chapter's Specifications.

14   If the lining material of the friction plates smells burnt or if it's glazed, new parts are required. If the steel clutch plates are scored or discolored, they must be replaced with new ones. Measure the thickness of each friction plate (see illustration) and compare the results to this chapter's Specifications. Replace the friction plates as a set if any are near the wear limit.

15   Lay the metal plates, one at a time, on a perfectly flat surface (such as a piece of plate glass) and check for warpage by trying to slip a gauge between the flat surface and the plate (see illustration). The feeler gauge should be the same thickness as the warpage limit listed in this chapter's Specifications. Do this at several places around the plate's circumference. If the feeler gauge can be slipped under the plate, it is warped and should be replaced with a new one.

16   Check the tabs on the friction plates for excessive wear and mushroomed edges. They can be cleaned up with a file if the deformation is not severe.

17   Check the edges of the slots in the clutch housing for indentations made by the friction plate tabs. If the indentations are deep they can prevent clutch release, so the housing should be replaced with a new one. If the indentations can be removed easily

with a file, the life of the housing can be prolonged to an extent.

18   Check the clutch pressure plate, clutch boss wire clip, clutch boss spring and seat plate for wear and damage if they were removed). Replace any worn or damaged parts.

19   Check the release bearing (see illustration) and clutch housing thrust plate for wear, damage or roughness. Replace them if their condition is uncertain or obviously bad. Check the bearing surface in the clutch housing and replace the clutch housing if it's worn or damaged. Inspect the slots in the clutch housing and the drive dogs in the oil pump gear and replace any worn or damaged parts.

20   Clean all traces of old gasket material from the clutch cover and its mating surface on the crankcase.

21   Check the clutch release for smooth operation (see illustration). Disassemble it to clean and lubricate or replace components as necessary.

Installation

Refer to illustrations 19.26a, 19.26b, 19.29, 19.30a, 19.30b and 19.31

22   Install the oil pump drive gear on the clutch housing, using a new snap-ring.

23   Coat the inside of the clutch housing with engine oil and install it on the transmission shaft. Coat the thrust washer with engine oil, then install the thrust washer and clutch boss. Install a new clutch boss lockwasher and the nut. Hold the clutch boss with one of the methods used in

19.21 If the release shaft is worn or damaged, remove the snap-ring and remove it from the clutch cover

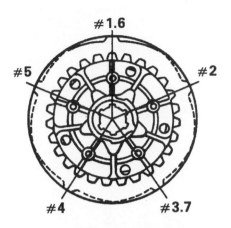

19.26a Position the tabs on the metal plate at the locations shown (plates 1 and 6 go at the top, 3 and 7 go at the lower right and so on)

19.26b The tabs on the metal plates align with the marks on the clutch boss (arrows)

19.29 Align the dots on the pressure plate and clutch boss (arrows)

Step 9, then tighten the nut to the torque listed in this Chapter's Specifications. Bend the lockwasher against the nut to secure it.

24   Install the clutch damper components (seat, clutch boss spring, clutch plate and wire circlip) if they were removed. Be sure both ends of the wire circlip fit into the hole in the clutch hub.

25   Coat one of the friction plates with engine oil and install it on the clutch housing. Engage the tabs on the friction plate with the slots in the clutch housing.

26   Install a metal plate on top of the friction plate. Note that each metal plate has a tab on the outer edge. The tabs should be staggered evenly as the metal plates are installed, lining them up with the marks on the clutch boss **(see illustrations).**

27   Coat the remaining friction plates with engine oil, then install alternating friction plates and metal plates until they're all installed (don't forget to stagger the metal plate tabs).

28   Coat the release bearing with oil and install the plate washer, bearing and pull rod in the pressure plate **(see illustration 19.7).** Make sure the gear teeth on the pull rod will face the rear of the engine when the pressure plate is installed.

29   Install the pressure plate, aligning the marks on pressure plate and clutch boss **(see illustration).**

30   Install the clutch springs, washers and pressure plate bolts **(see illustrations)**, then tighten the bolts evenly in a criss-cross pattern to the torque listed in this Chapter's Specifications.

19.30a Install the clutch springs, washers and bolts . . .

19.30b ... make sure the pull rod gear teeth face the rear of the engine (arrow), then tighten the bolts to Specifications in a criss-cross pattern

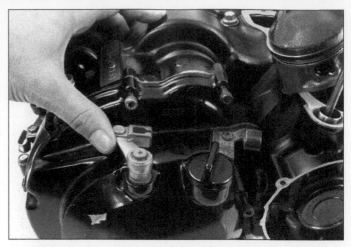

19.31 Push in on the release lever to position the clutch cover on the engine

31   Make sure the clutch cover dowels are in position and install a new gasket. Push the clutch release lever into position on the cover for tightening **(see illustration)**.
32   Tighten the clutch cover bolts evenly to the torque listed in this Chapter's Specifications.
33   Fill the crankcase with the recommended type and amount of engine oil (see Chapter 1).
34   The remainder of installation is the reverse of the removal steps.

20 Clutch cable - removal and installation

Removal

Refer to illustrations 20.2, 20.3 and 20.4
1   Remove body panels as necessary for access to the lower clutch cable nuts on the right side of the engine.
2   Loosen the clutch cable nuts **(see illustration)**. Disconnect the lower end of the cable from the release lever and the bracket on the engine.
3   Pull back the rubber cover from the clutch adjuster at the handlebar **(see illustration)**. Loosen the lockwheel, then the adjusting nut to create slack in the cable.
4   Pull the adjuster out of its slot in the clutch lever bracket, then

20.2 Loosen the nuts and detach the clutch cable from the bracket

rotate the cable to align it with the slot in the clutch lever and lower the cable end out of the clutch lever **(see illustration)**.
5   Installation is the reverse of the removal steps. Adjust the clutch lever freeplay (see Chapter 1).

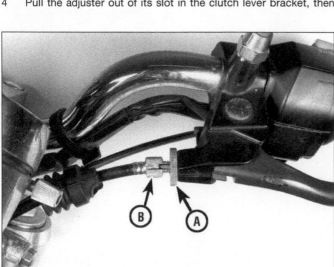

20.3 Loosen the locknut (A) and turn the adjuster (B) to create slack in the cable

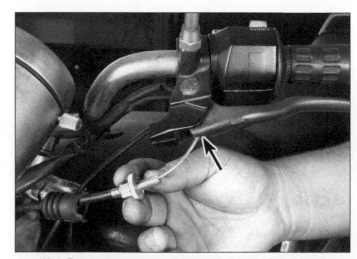

20.4 Rotate the cable to align it with the slot (arrow) and lower it out of the clutch lever

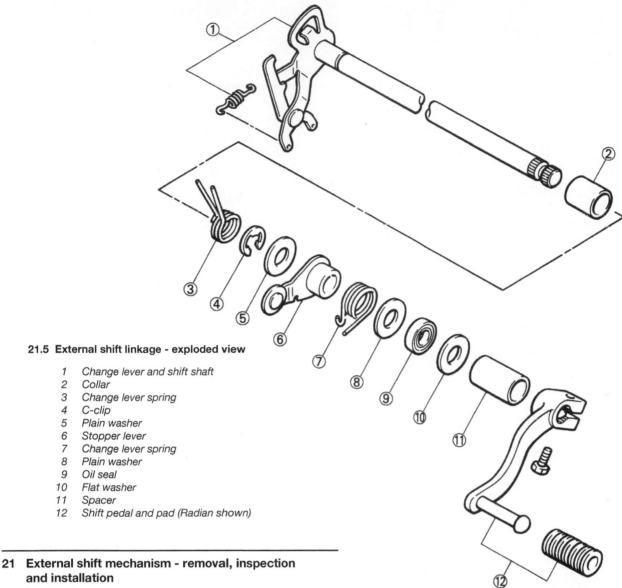

21.5 External shift linkage - exploded view

1   Change lever and shift shaft
2   Collar
3   Change lever spring
4   C-clip
5   Plain washer
6   Stopper lever
7   Change lever spring
8   Plain washer
9   Oil seal
10 Flat washer
11 Spacer
12 Shift pedal and pad (Radian shown)

21 External shift mechanism - removal, inspection and installation

Shift linkage and pedal

Refer to illustration 21.5

1   FJ/XJ600 models and FZ600 models use a shift pedal and linkage rod connected to the shift shaft. YX600 Radian models use a shift pedal attached directly to the shift shaft.

2   Set the bike on its centerstand (if equipped) or prop it securely upright.

3   Where necessary for access, remove the left footpeg bracket.

4   Look for alignment marks on the end of the shift pedal or lever and the shift shaft. If they aren't visible, make your own.

5   Remove the pinch bolt and detach the shift pedal (YX600 Radian) or shift lever (all others) from the shift shaft **(see illustration).**

6   Installation is the reverse of the removal steps. Tighten the shift pedal bolt to the torque listed in this Chapter's Specifications.

Shift mechanism removal

Refer to illustrations 21.9 and 21.11

7   Remove the shift pedal or linkage (see Steps 1 through 5 above).

8   Remove the engine sprocket cover (see Chapter 5).

9   Remove the plastic spacer and flat washer from the shift shaft **(see illustration 21.5 and the accompanying illustration).**

21.9 Remove the plastic spacer and flat washer; check the condition of the seal (arrow)

21.11 Pull the change lever (A) away from the shift drum and slide the shift shaft (B) out of the crankcase; the change lever spring finger should be positioned on either side of the guide bar (C) when the shift shaft is installed

22.5 Remove the screws (arrows) and take the cover off the crankcase

22.6 Bend back the lockwasher, unscrew the nut and remove the primary gear

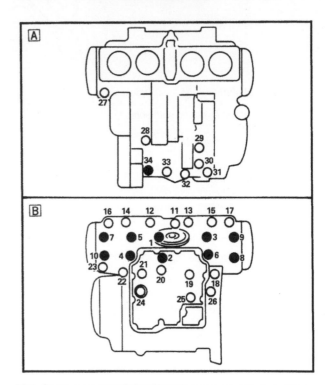

22.7 Crankcase bolt TIGHTENING sequence; 8mm bolts are marked in black and 6 mm bolts are marked in white

10   Remove the clutch (see Section 19).
11   Pull the change lever away from the shift cam and slide the shift shaft and washer out of the crankcase **(see illustration)**. Slip the flat washer and plastic spacer back onto the shaft so they can be kept in order **(see illustration 21.5)**.

Shift mechanism inspection

12   Inspect the shift shaft return springs **(see illustration 21.5)**. If they are worn or damaged, replace them.
13   Check the change shaft for bends and damage to the splines. If the shaft is bent, you can attempt to straighten it, but if the splines are damaged it will have to be replaced. Inspect the pawl and springs on the change shaft and replace the shaft if they're worn or damaged.
14   Check the condition of the stopper lever and spring. Replace the stopper lever if it's worn where it contacts the shift cam. Replace the spring if it's distorted.
15   Inspect the straight detent pins on the end of the shift cam. If they're worn or damaged, you'll have to disassemble the crankcase to replace the shift cam.
16   Check the condition of the seal on the right side of the engine case. If it has been leaking, pry it out **(see illustration 21.9)**. It's a good idea to replace it in any case, since gaining access to it requires a fair amount of work. Install a new seal with its closed side facing outward.

It should be possible to install the seal with thumb pressure, but if necessary, drive it in with a socket the same diameter as the seal.

Shift mechanism installation

17   Remove the circlip and flat washer from the end of the shift shaft. Make sure the large washer is still on the shaft, positioned against the change lever.
18   Apply high-temperature grease to the lip of the seal. Wrap the splines of the shift shaft with electrical tape, so the splines won't damage the seal as the shaft is installed.
19   Slide the shaft into the crankcase. Engage the pawls with the pins on the shift cam and position the return spring over its guide bar **(see illustration 21.11)**.
20   The remainder of installation is the reverse of the removal steps.
21   Refill the engine oil (see Chapter 1).

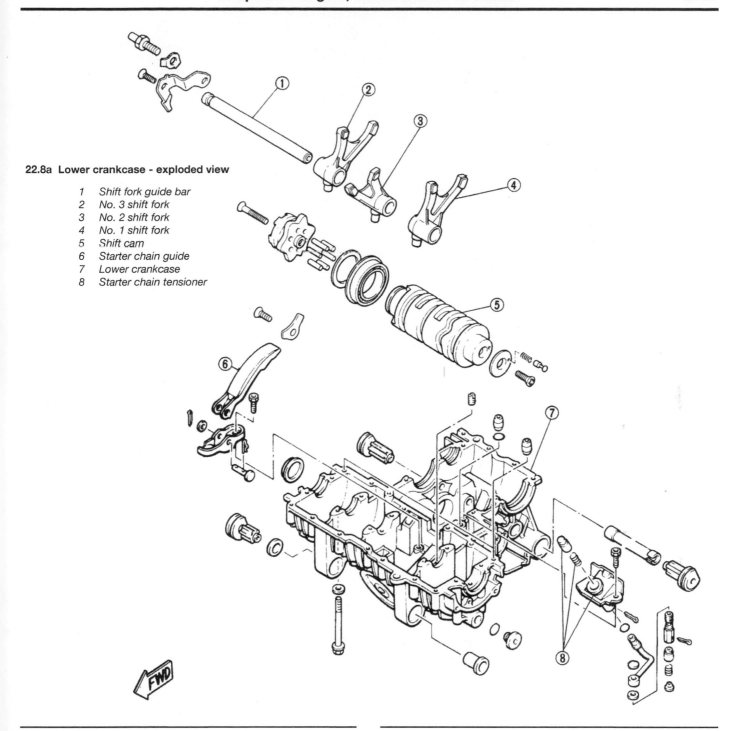

22.8a Lower crankcase - exploded view

1 Shift fork guide bar
2 No. 3 shift fork
3 No. 2 shift fork
4 No. 1 shift fork
5 Shift cam
6 Starter chain guide
7 Lower crankcase
8 Starter chain tensioner

22 Crankcase - disassembly and reassembly

1 To examine and repair or replace the crankshaft, starter and camshaft chains, connecting rods, bearings, transmission components, starter idle gears or starter motor clutch, the crankcase must be split into two parts.

Disassembly

Refer to illustrations 22.5, 22.6, 22.7, 22.8a and 22.8b

2 If the crankcase is being separated to remove the crankshaft, remove the cylinder head, cylinder block and pistons (see Sections 10, 13 and 14). If you're only separating the crankcase halves to disassemble the transmission shafts or remove the internal shift linkage, there's no need to remove the top-end components. In all cases, remove

the clutch (see Section 19).

3 Remove the oil pan (see Section 16). Remove the relief valves from the crankcase.

4 Remove the signal generator and alternator rotors (see Chapters 4 and 8).

5 Remove the cover and gasket from the right front corner of the crankcase **(see illustration)**.

6 Check the primary gear for wear or damage **(see illustration)**. If necessary, bend back the tabs of the lockwasher and remove the nut, then slide the gear off.

7 Remove the upper crankcase bolts, then the lower crankcase bolts, starting with the highest-numbered bolt and working to the lowest **(see illustration)**.

8 Carefully separate the crankcase halves **(see illustrations)**. If

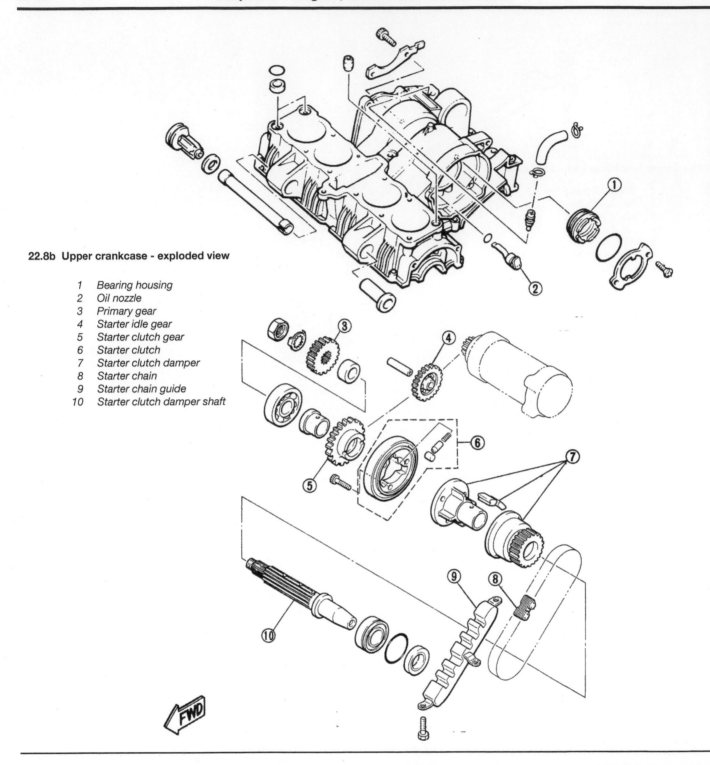

22.8b Upper crankcase - exploded view

 1   *Bearing housing*
 2   *Oil nozzle*
 3   *Primary gear*
 4   *Starter idle gear*
 5   *Starter clutch gear*
 6   *Starter clutch*
 7   *Starter clutch damper*
 8   *Starter chain*
 9   *Starter chain guide*
10   *Starter clutch damper shaft*

they won't come apart easily, make sure all fasteners have been removed. Don't pry against the crankcase mating surfaces or they will leak after reassembly.

9   Refer to Sections 23 through 31 for information on the internal components of the crankcase.

Reassembly

Refer to illustration 22.15

10   Remove all traces of sealant from the crankcase mating surfaces. Be careful not to let any fall into the case as this is done. Check to make sure the O-rings and dowel pins are in place in their holes in the mating surface of the lower crankcase half **(see illustrations 22.8a and 22.8b)**.

11   A common method of assembling motorcycle crankcases is to position the engine upside down and install the crankshaft and transmission shafts in it, then install the lower crankcase half on top of the upper crankcase half. Yamaha recommends positioning the upper crankcase upside down with the crankshaft in it, but installing the transmission shafts in the lower crankcase half and holding them in place with a suitable bar while lowering the lower crankcase half into the upper crankcase half.

12   Pour some engine oil over the transmission gears, the crankshaft main bearings and the shift cam. don't get any oil on the crankcase mating surfaces.

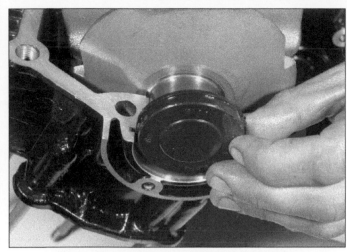

22.15 Make sure the blind plug is in place before assembling the crankcase

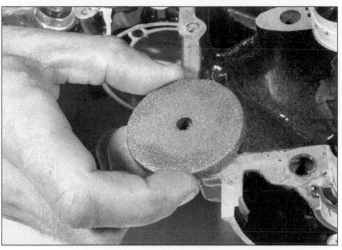

23.3 Small burrs can be removed from the gasket surfaces with a fine sharpening stone or grindstone

13   Apply a thin, even bead of Yamabond 4 Quick Gasket sealant or equivalent to the crankcase mating surfaces. **Caution**: *don't apply an excessive amount of sealant. don't let it contact the oil gallery O-ring. don't apply it within 2 to 3 mm (5/64 to 1/8 inch) of the bearing inserts, as it will ooze out when the case halves are assembled and may obstruct oil passages and prevent the bearings from seating.*

14   Check the position of the shift cam, shift forks and transmission shafts - make sure they're in the neutral position. When this is the case, it will be possible to rotate each pair of gears independently of the other gears on the shafts.

15   Carefully assemble the crankcase halves over the dowels and the blind plug at the end of the crankshaft **(see illustration)**. While doing this, make sure the shift forks fit into their gear grooves. Make sure the crankshaft blind plug is positioned correctly in the case grooves). Check the input and output shaft bearings to ensure that the bearing positioners are installed correctly. **Caution**: *the crankcase halves should fit together completely without being forced. If they're slightly apart, DO NOT force them together by tightening the crankcase bolts. The most likely reason they're apart is that the transmission bearing pins aren't aligned with their holes. If the pins are forced against the crankcase halves, the cases will crack and have to be replaced.*

16   Install the crankcase bolts (and washers, if equipped) in the correct holes **(see illustration 22.7)**. Bolt 26 secures a wiring harness retainer. Bolt 32 secures a ground (earth) wire.

17   Install the oil pan (see Section 16).

18   Turn the transmission shafts to make sure they turn freely. Also make sure the crankshaft turns freely.

19   The remainder of installation is the reverse of removal, with the following additions:

a) *Once the external shift linkage is installed, shift the transmission through all the gear positions and back to Neutral.*

b) *Use a new lockwasher on the primary gear (if removed) and tighten its nut to the torque listed in this Chapter's Specifications.*

c) *Be sure to refill the engine oil with the grade listed in Chapter 1.*

23 Crankcase components - inspection and servicing

Refer to illustration 23.3

1   After the crankcases have been separated and the crankshaft, shift cam and forks and transmission components removed, the crankcases should be cleaned thoroughly with new solvent and dried with compressed air.

2   Remove any oil passage plugs that haven't already been removed. All oil passages should be blown out with compressed air.

3   All traces of old gasket sealant should be removed from the

mating surfaces. Minor damage to the surfaces can be cleaned up with a fine sharpening stone or grindstone **(see illustration)**. **Caution**: *Be very careful not to nick or gouge the crankcase mating surfaces or leaks will result. Check both crankcase halves very carefully for cracks and other damage.*

4   Check the cam chain guides for wear (see Section 27). If they appear to be worn excessively, replace them.

5   If any damage is found that cannot be repaired, replace the crankcase halves as a set.

24 Main and connecting rod bearings - general note

1   Even though main and connecting rod bearings are generally replaced with new ones during the engine overhaul, the old bearings should be retained for close examination as they may reveal valuable information about the condition of the engine.

2   Bearing failure occurs mainly because of lack of lubrication, the presence of dirt or other foreign particles, overloading the engine and/or corrosion. Regardless of the cause of bearing failure, it must be corrected before the engine is reassembled to prevent it from happening again.

3   When examining the bearings, remove the main bearings from the case halves and the rod bearings from the connecting rods and caps and lay them out on a clean surface in the same general position as their location on the crankshaft journals. This will enable you to match any noted bearing problems with the corresponding crankshaft journal.

4   Dirt and other foreign particles get into the engine in a variety of ways. It may be left in the engine during assembly or it may pass through filters or breathers. It may get into the oil and from there into the bearings. Metal chips from machining operations and normal engine wear are often present. Abrasives are sometimes left in engine components after reconditioning operations such as cylinder honing, especially when parts are not thoroughly cleaned using the proper cleaning methods. Whatever the source, these foreign objects often end up imbedded in the soft bearing material and are easily recognized. Large particles will imbed in the bearing and will score or gouge the bearing and journal. The best prevention for this cause of bearing failure is to clean all parts thoroughly and keep everything spotlessly clean during engine reassembly. Frequent and regular oil and filter changes are also recommended.

5   Lack of lubrication or lubrication breakdown has a number of interrelated causes. Excessive heat (which thins the oil), overloading (which squeezes the oil from the bearing face) and oil leakage or throw off (from excessive bearing clearances, worn oil pump or high engine speeds) all contribute to lubrication breakdown. Blocked oil passages will also starve a bearing and destroy it. When lack of lubrication is the

25.3a Remove the starter clutch and lift the crankshaft out, together with the connecting rods, starter chain and cam chain

cause of bearing failure, the bearing material is wiped or extruded from the steel backing of the bearing. Temperatures may increase to the point where the steel backing and the journal turn blue from overheating.

6   Riding habits can have a definite effect on bearing life. Full throttle low speed operation, or lugging (labouring) the engine, puts very high loads on bearings, which tend to squeeze out the oil film. These loads cause the bearings to flex, which produces fine cracks in the bearing face (fatigue failure). Eventually the bearing material will loosen in pieces and tear away from the steel backing. Short trip driving leads to corrosion of bearings, as insufficient engine heat is produced to drive off the condensed water and corrosive gases produced. These products collect in the engine oil, forming acid and sludge. As the oil is carried to the engine bearings, the acid attacks and corrodes the bearing material.

7   Incorrect bearing installation during engine assembly will lead to bearing failure as well. Tight fitting bearings which leave insufficient bearing oil clearances result in oil starvation. Dirt or foreign particles trapped behind a bearing insert result in high spots on the bearing which lead to failure.

8   To avoid bearing problems, clean all parts thoroughly before reassembly, double check all bearing clearance measurements and lubricate the new bearings with engine assembly lube or moly-based grease during installation.

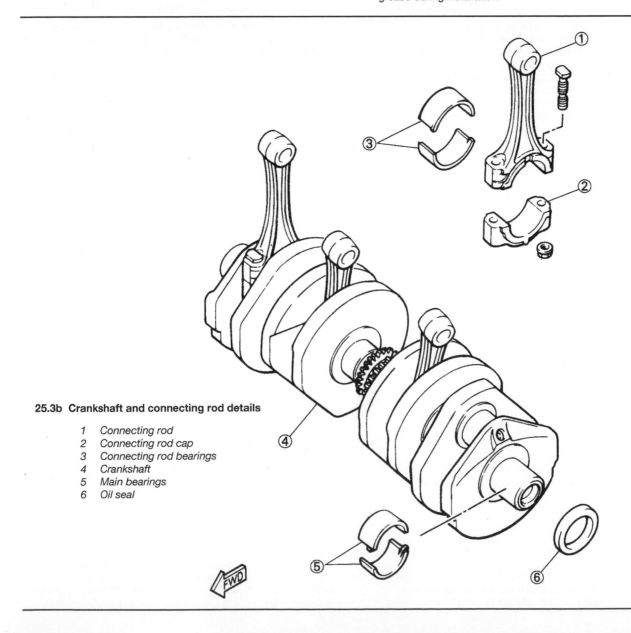

25.3b Crankshaft and connecting rod details

1   *Connecting rod*
2   *Connecting rod cap*
3   *Connecting rod bearings*
4   *Crankshaft*
5   *Main bearings*
6   *Oil seal*

25.4 Remove the bearing inserts

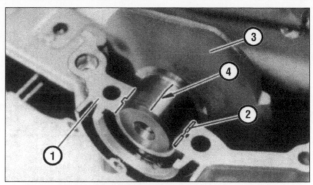

25.9 Lay the Plastigage strips on the journals, parallel to the crankshaft centerline

1	Crankcase	3	Crankshaft
2	Bearing	4	Plastigage

25 Crankshaft and main bearings - removal, inspection, main bearing selection and installation

Crankshaft removal

Refer to illustrations 25.3a, 25.3b and 25.4

1   Before removing the crankshaft check the endplay, using a dial indicator mounted in line with the crankshaft. Yamaha doesn't provide endplay specifications, but if the endplay is excessive (more than about 0.15 mm or 0.006"), consider replacing the case halves.

2   Remove the starter clutch damper (see Section 31).

3   Lift the crankshaft out, together with the connecting rods, cam chain and starter chain and set them on a clean surface **(see illustrations)**.

4   The main bearing inserts can be removed from their saddles by pushing their centers to the side, then lifting them out **(see illustration)**. Keep the bearing inserts in order. The main bearing oil clearance should be checked, however, before removing the inserts.

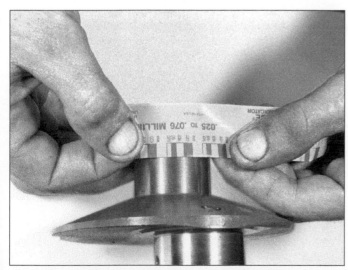

25.11 Measuring the width of the crushed Plastigage (be sure to use the correct scale - standard and metric are included)

Inspection

5   If you Haven't already done so, mark and remove the connecting rods from the crankshaft (see Section 26).

6   Clean the crankshaft with solvent, using a rifle-cleaning brush to scrub out the oil passages. If available, blow the crank dry with compressed air. Check the main and connecting rod journals for uneven wear, scoring and pits. Rub a copper coin across the journal several times - if a journal picks up copper from the coin, It's too rough. Replace the crankshaft.

7   Check the camshaft chain sprocket and starter chain sprocket on the crankshaft for chipped teeth and other wear. If any undesirable conditions are found, replace the crankshaft. Check the chains as described in Section 27. Check the rest of the crankshaft for cracks and other damage. It should be Magnafluxed to reveal hidden cracks - a dealer service department or motorcycle machine shop should handle this procedure.

8   Set the crankshaft on V-blocks and check the runout with a dial indicator touching each of the main journals, comparing your findings with this Chapter's Specifications. If the runout exceeds the limit, replace the crank.

Main bearing selection

Refer to illustrations 25.9, 25.11, 25.14 and 25.16a through 25.16e

9   To check the main bearing oil clearance, clean off the bearing inserts (and reinstall them, if they've been removed from the case) and lower the crankshaft into the upper half of the case. Cut four pieces of Plastigage (type HPG-1) and lay them on the crankshaft main journals, parallel with the journal axis **(see illustration)**.

10   Very carefully, guide the lower case half down onto the upper case half. *Install the crankshaft retaining bolts (case bolts 1 through 10) and tighten them, using the recommended sequence, to the torque listed in this chapter's Specifications* (see Section 22). **Caution**: *DO NOT tighten the bolts unless the case halves fit together completely. don't rotate the crankshaft during this procedure!*

11   Now, remove the bolts and carefully lift the lower case half off. Compare the width of the crushed Plastigage on each journal to the scale printed on the Plastigage envelope to obtain the main bearing oil clearance **(see illustration)**. Write down your findings, then remove all traces of Plastigage from the journals, using your fingernail or the edge of a credit card.

12   If the oil clearance falls into the specified range, no bearing replacement is required (provided they are in good shape). If the clearance is more than the standard range, but within the service limit, replace the bearing inserts with new ones of the same thickness and check the oil clearance once again. Replace all of the inserts, as a set, at the same time.

13   The clearance should be within the range listed in this chapter's Specifications.

25.14 Measure the diameter of each crankshaft journal at several points to detect taper and out-of-round conditions

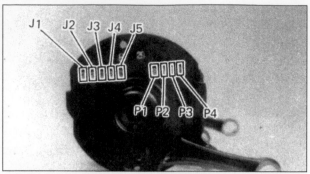

25.16a These numbers on the crankshaft indicate journal thickness; reading from left to right, the first five numbers correspond with main bearing journals no. 1 through 5 - the next four numbers correspond with connecting rod journals no. 1 through 4

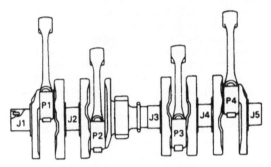

25.16b The main and rod journals are numbered from the left to the right side of the engine

25.16c The numbers on the crankcase correspond with the crankshaft main journals no. 1 through 5

Bearing color code	
No. 1	Blue
No. 2	Black
No. 3	Brown
No. 4	Green
* No. 5	Yellow

* No. 5 applies only to the crankshaft main bearing selection.

25.16d Calculate the bearing number by subtracting the crankshaft number from the crankcase number, then use the bearing number to select a color code

25.16e The color codes painted on the sides of the bearings identify bearing thickness

14   If the clearance is greater than the service limit listed in this chapter's Specifications, measure the diameter of the crankshaft journals with a micrometer (see illustration). Yamaha doesn't provide journal diameter or wear specifications, but by measuring the diameter at a number of points around each journal's circumference, you'll be able to determine whether or not the journal is out-of-round. Take the measurement at each end of the journal, near the crank throws, to determine if the journal is tapered. Out-of-round and taper should not exceed 0.04 millimeters (0.001").

15   If any crank journal is out-of-round or tapered or the bearing clearance is beyond the limit listed in this chapter's Specifications with new bearings, replace the crankshaft and bearings as a complete set.

16   Use the number marks on the crankshaft and on the case to determine the bearing sizes required. The first five numbers on the crankshaft web are the main journal numbers, starting with the left journal (see illustrations). These correspond with the numbers at the rear of the upper crankcase half (see illustration). To determine the bearing number for each bearing, subtract the crankshaft number from the case number. For example, the crankshaft number for journal no. 1 is 2 and the case number for journal no. 1 is 6. Subtracting 2 from 6 produces 4, which is the bearing number for journal no. 1. According to the accompanying chart, bearing no. 4 is color-coded green (see illustration). The color codes are painted on the edges of the bearings (see illustration).

Installation

Refer to illustrations 25.17a and 25.17b

17   Clean the bearing saddles in the case halves, then install the bearing inserts in their saddles in the case (see illustrations). When installing the bearings, use your hands only - don't tap them into place with a hammer! Do not get finger prints on the front or rear of the bearings.

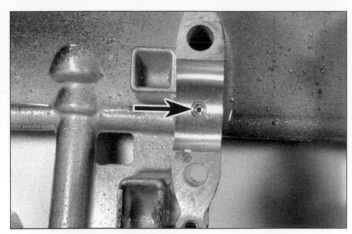

25.17a Make sure the oil holes are clear (arrow) . . .

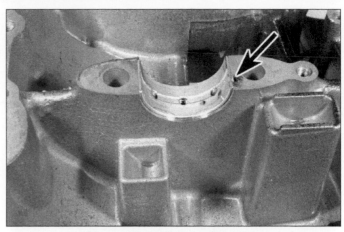

25.17b . . . then install the bearings in their saddles - engage the locating tab securely in its notch (arrow)

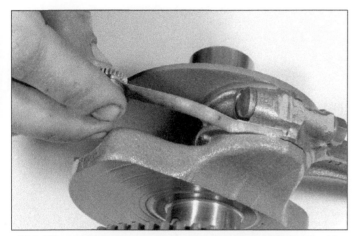

26.1 Measure the clearance between the crankshaft and the big end of the connecting rod with a feeler gauge

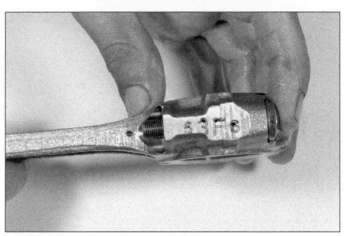

26.2 Make a cylinder number mark on the connecting rod; the letter stamped across the rod cap is a guide for reassembly; the number 3 or 4 is used in bearing selection

22 Carefully lower the crankshaft into place. If the connecting rods are in the engine, guide them onto the crankshaft journals and refit their caps (see Section 26).

23 Assemble the case halves (see Section 22) and check to make sure the crankshaft and the transmission shafts turn freely.

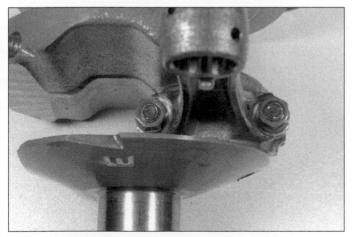

26.3a Remove the connecting rod nuts . . .

18 Lubricate the bearing inserts with engine assembly lube or moly-based grease.

19 You can install the connecting rods on the crankshaft at this point if the top end was removed from the engine (see Section 26).

20 Loop the camshaft chain and the starter chain over the crankshaft and lay them onto their sprockets.

21 If the connecting rods are in the engine, place short pieces of hose over the studs to protect the crankshaft as you remove them.

26 Connecting rods and bearings - removal, inspection, bearing selection and installation

Removal

Refer to illustrations 26.1, 26.2, 26.3a and 26.3b

1 Before removing the connecting rods from the crankshaft, measure the side clearance of each rod with a feeler gauge (see illustration). If the clearance on any rod is greater than that listed in this chapter's Specifications, that rod may have to be replaced with a new one. Check the crankshaft journal for excessive, if excessive wear is evident, replace the crankshaft and/or the connecting rods as a set.

2 Use a center punch or number stamp to mark the position of each rod and cap, relative to its position on the crankshaft (see illustration). The printed numbers on the rods don't indicate cylinder number; they're used for bearing selection.

3 Unscrew the bearing cap nuts, separate the cap from the rod, then detach the rod from the crankshaft (see illustrations). If the cap is stuck, tap on the ends of the rod bolts with a soft face hammer to free them.

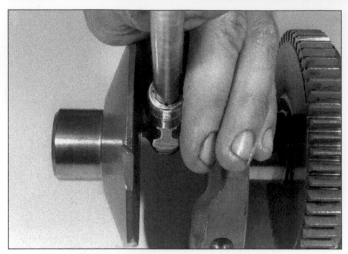

26.3b . . . with a socket and extension

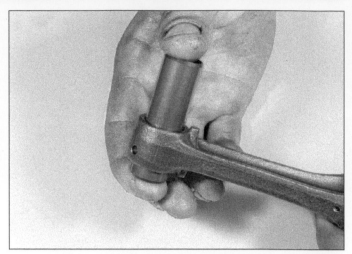

26.5 Slip the piston pin into the rod and rock it back-and-forth to check for looseness

4   Separate the bearing inserts from the rods and caps, keeping them in order so they can be reinstalled in their original locations. Wash the parts in solvent and dry them with compressed air, if available.

Inspection

Refer to illustration 26.5

5   Check the connecting rods for cracks (this may require taking the rods to a machine shop) and other obvious damage. Lubricate the piston pin for each rod, install it in the proper rod and check for play **(see illustration)**. If it wobbles, replace the connecting rod and/or the pin.

6   Refer to Section 26 and examine the connecting rod bearing inserts. If they are scored, badly scuffed or appear to have been seized, new bearings must be installed. Always replace the bearings in the connecting rods as a set. If they're badly damaged, check the corresponding crankshaft journal. Evidence of extreme heat, such as discoloration, indicates that lubrication failure has occurred. Be sure to thoroughly check the oil passages, oil pump and pressure relief valves as well as all oil holes and passages for debris before reassembling the engine.

7   Have the rods checked for twist and bend at a dealer service department or motorcycle repair shop.

Connecting rod bearing selection

Refer to illustrations 26.13 and 26.18

8   If the bearings and journals appear to be in good condition, check the oil clearances as follows:

9   Start with the rod for the no. 1 cylinder. Wipe the bearing inserts and the connecting rod and cap clean, using a lint-free cloth.

10   Install the bearing inserts in the connecting rod and cap. Make sure the tab on the bearing engages with the notch in the rod or cap.

11   Wipe off the connecting rod journal with a lint-free cloth. Lay a strip of Plastigage (type HPG-1) across the top of the journal, parallel with the journal axis **(see illustration 25.9)**.

12   Position the connecting rod on the bottom of the journal, then install the rod cap and nuts. Tighten the nuts to the torque listed in this chapter's Specifications, but don't allow the connecting rod to rotate at all.

13   Unscrew the nuts and remove the connecting rod and cap from the journal, being very careful not to disturb the Plastigage. Compare the width of the crushed Plastigage to the scale printed in the Plastigage envelope **(see illustration)** to determine the bearing oil clearance.

14   If the clearance is within the range listed in this chapter's Specifications and the bearings are in perfect condition, they can be reused. If the clearance is beyond the standard range, replace the

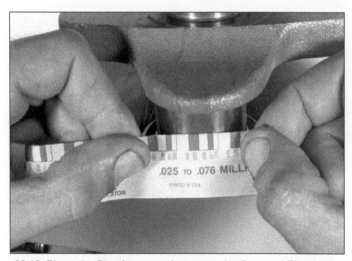

26.13 Place the Plastigage scale next to the flattened Plastigage to measure the bearing clearance

bearing inserts with new inserts that have the same color code, then check the oil clearance once again. Always replace all of the inserts at the same time.

15   The clearance should be within the range listed in this chapter's Specifications.

16   If the clearance is greater than the maximum clearance listed in this chapter's Specifications, measure the diameter of the connecting rod journal with a micrometer. As with the main bearing journals, Yamaha doesn't provide diameter or wear limit specifications, but by measuring the diameter at a number of points around the journal's circumference, you'll be able to determine whether or not the journal is out-of-round. Take the measurement at each end of the journal to determine if the journal is tapered.

17   If any journal is tapered or out-of-round or bearing clearance is beyond the maximum listed in this chapter's Specifications, replace the crankshaft.

18   Each connecting rod has a 3 or 4 stamped on it in ink **(see illustration 26.2)**. Subtract this number from the connecting rod journal number on the crankshaft to get a bearing number **(see illustrations 25.16a and 25.16b)**. For example, the number on the connecting rod shown in illustration 26.2 is 3. The corresponding number for that connecting rod's journal, stamped into the crankshaft, is 2. Subtracting 2 from 3 produces 1, which is the bearing number for that journal. According to the chart in illustration 25.16d, bearing no. 1

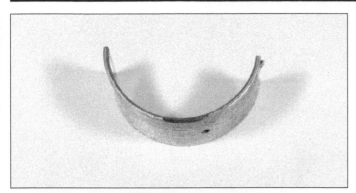

26.18 The color code is painted on the side of the bearing

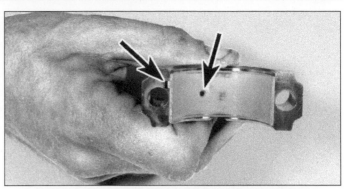

26.20 Be sure the tab fits in the notch and the oil hole in the upper bearing aligns with the hole in the connecting rod (arrows)

26.21 The Y mark on the rod faces the left side of the engine

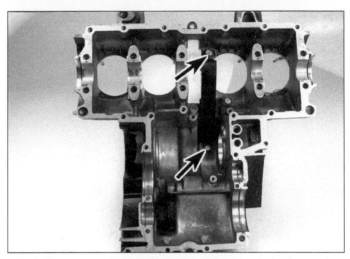

27.8 The starter chain guide is secured by two bolts (arrows); use non-permanent thread locking agent on the bolts during installation

is color-coded blue. The color codes are painted on the edges of the bearings **(see illustration)**.

19   Repeat the bearing selection procedure for the remaining connecting rods.

Installation

Refer to illustrations 26.20 and 26.21

20   Wipe off the bearing inserts, connecting rods and caps. Install the inserts into the rods and caps, using your hands only, making sure the tabs on the inserts engage with the notches in the rods and caps and the oil holes in rod and bearing line up **(see illustration)**. When all the inserts are installed, lubricate them with engine assembly lube or moly-based grease. don't get any lubricant on the back side of the bearings in the rod or cap.

21   Assemble each connecting rod to its proper journal, referring to the previously applied cylinder numbers. Make sure the Y mark on the rod is toward the left side of the engine **(see illustration)**. Also, the letter present at the rod/cap seam on one side of the connecting rod should fit together perfectly when the rod and cap are assembled **(see illustration 26.2)**. If it doesn't, the wrong cap is on the rod. Fix this problem before assembling the engine any further.

22   When you're sure the rods are positioned correctly, lubricate the threads of the rod bolts with molybdenum disulphide grease and tighten the nuts to the torque listed in this chapter's Specifications. **Note**: *Tighten the nuts in a continuous motion. If you must stop before the nuts are fully tightened, loosen them completely, then retighten them to the specified torque.*

23   Turn the rods on the crankshaft. If any of them feel tight, tap on the bottom of the connecting rod caps with a soft hammer - this

should relieve stress and free them up. If it doesn't, recheck the bearing clearance.

24   As a final step, recheck the connecting rod side clearances (see Step 1). If the clearances aren't correct, find out why before proceeding with engine assembly.

27   Starter chain, camshaft chain and guides - removal, inspection and installation

Removal

Starter chain and camshaft chain

1   Remove the engine (see Section 5).
2   Separate the crankcase halves (see Section 22).
3   Remove the starter drive clutch shaft (see Section 31).
4   Remove the crankshaft (see Section 25).
5   Remove the chains from the crankshaft.

Chain guides

Refer to illustration 27.8

6   The exhaust side (front) cam chain guide can be lifted from the cylinder head **(see illustration 9.6b)**.

7   The cam chain rear guide is held in position by a pushrod, spring and bolt **(see illustration 8.3)**. Remove the cylinder block (see Section 13), then remove the bolt, washer, spring and damper pushrod from the crankcase. Lift out the guide.

8   The starter chain guide in the lower case half is secured by two bolts **(see illustration)**.

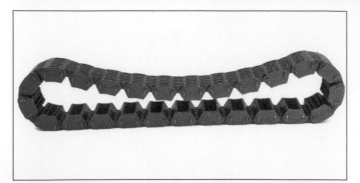

27.9 Replace the starter chain if it's worn or damaged

28.2a Lift out the mainshaft (arrow) . . .

28.2b . . . and the driveshaft

28.4 Be sure the bearing retainer half-rings are in their slots (arrows); there's one at the end of each shaft

Inspection

Refer to illustration 27.9

Starter chain and camshaft chain

9   Check the chains for binding and obvious damage. Lay the starter drive chain onto a flat surface. The inner edges of the chain should not touch **(see illustration)**. If this condition is visible, or if the chain appears to be stretched, replace it. The camshaft drive chain should be checked for stretch and to see if the rollers are in good working order.

Chain guides

10   Check the guides for deep grooves, cracking and other obvious damage, replacing them if necessary.

Installation

11   Installation of these components is the reverse of the removal procedure, with the following additions:

a) *When installing the starter chain guide, apply a non-hardening thread locking compound to the threads of the bolts. Tighten the bolts to the torque listed in this chapter's Specifications.*
b) *Apply engine oil to the faces of the guides and to the chains.*
c) *The rear cam chain guide components can be installed from inside the crankcase if the halves are separated. Be sure the pushrod and guide engage each other.*
d) *When installing the cam chain guides, be sure they rest in their sockets in the lower crankcase half.*
e) *Use a new sealing washer on the exhaust side cam chain guide bolt. Tighten the bolt securely, but take care not to strip out the threads in the case.*

28   Transmission shafts - removal and installation

Refer to illustrations 28.2a and 28.2b

Removal

1   Remove the engine and clutch, then separate the case halves (see Sections 5, 20, and 22).
2   Lift out the mainshaft, then the driveshaft **(see illustrations)**. If they are stuck, use a soft-face hammer and gently tap on the bearings on the ends of the shafts to free them. Be careful not to lose the bearing end caps and bearing retainer half-rings.
3   Refer to Section 29 for information pertaining to transmission shaft service and Section 30 for information pertaining to the shift cam and forks.

Installation

Refer to illustration 28.4

4   Carefully lower each shaft into place, making sure the shift forks engage the grooves in the gears **(see illustrations 28.2a and 28.2b)**. The bearing end caps should be held in place with a little moly lube. The bearing retainer half-rings should fit into the locating grooves in the crankcase and the ball bearing outer races **(see illustrations)**. **Caution:** *if the rings are out of position and you try to force the crankcase halves together by tightening the bolts, the crankcase halves will crack and will have to be replaced.*

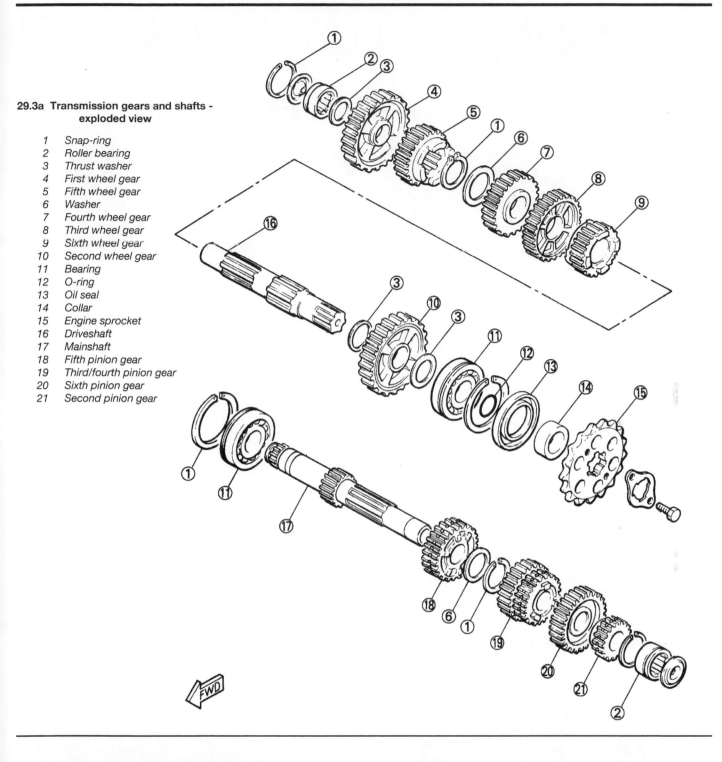

29.3a Transmission gears and shafts - exploded view

1 Snap-ring
2 Roller bearing
3 Thrust washer
4 First wheel gear
5 Fifth wheel gear
6 Washer
7 Fourth wheel gear
8 Third wheel gear
9 Sixth wheel gear
10 Second wheel gear
11 Bearing
12 O-ring
13 Oil seal
14 Collar
15 Engine sprocket
16 Driveshaft
17 Mainshaft
18 Fifth pinion gear
19 Third/fourth pinion gear
20 Sixth pinion gear
21 Second pinion gear

5   The remainder of installation is the reverse of removal.
6   Make sure the gears are in the neutral position. When this occurs, each pair of gears can be rotated independently of the others.

29 Transmission shafts - disassembly, inspection and reassembly

Note: *When disassembling the transmission shafts, place the parts on a long rod or thread a wire through them to keep them in order and facing the proper direction.*
1   Remove the shafts from the case (see Section 28).

2   The gear bushings are a press fit on the shafts, so a hydraulic press is required at several points to disassemble and reassemble the shafts. If you don't have one, it may be more convenient to have a dealer do the disassembly and reassembly.

Mainshaft

Disassembly

Refer to illustrations 29.3a, 29.3b, 29.4 and 29.5

3   Slide the bearing off the mainshaft and remove the snap-ring **(see illustrations)**.

29.3b Remove the bearing and snap-ring from the
end of the shaft

29.4 Press the shaft out of the sixth and second gears

4   Press the second and sixth pinion gears loose from the shaft together, then remove them from the shaft **(see illustration)**.

5   Slide the third and fourth gear off the shaft **(see illustration)**. Remove the snap-ring and thrust washer, then slide the fifth pinion gear off the shaft.

Inspection

Refer to illustrations 29.6 and 29.8

6   Wash all of the components in clean solvent and dry them off. Rotate the ball bearing on the shaft, feeling for tightness, rough spots and excessive looseness and listening for noises. If the bearing is defective, press it off the shaft **(see illustration).**

7   Inspect the roller bearings for pitting and scoring and replace it if its condition seems at all doubtful.

8   Check the gear teeth for cracking and other obvious damage. Check the gear bushings and the surface in the inner diameter of each gear for scoring or heat discoloration **(see illustration)**. If the gear or bushing is damaged, replace it. First pinion gear is integral with the mainshaft; the mainshaft must be replaced if the gear is defective.

9   Inspect the dogs and the dog holes in the gears for excessive wear. Replace the paired gears as a set if necessary.

10   Place the shaft in V-blocks and check runout with a dial indicator. Replace the shaft if runout exceeds the value listed in this chapter's Specifications.

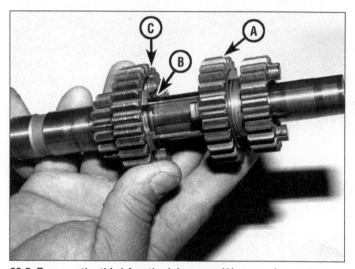

29.5 Remove the third-fourth pinion gear (A), snap-ring and thrust
washer (B) and the fifth pinion gear (C)

29.6 If the bearing needs to be replaced, press it off the shaft

29.8 Replace the gears if their internal bushings (arrow)
are worn or damaged

29.11a With fifth gear, its thrust washer and snap-ring and third-fourth gear on the shaft, press on sixth and second gears

29.11b Position the bearing so its snap-ring groove is away from the gears, then press it onto the shaft

29.11c The assembled gears and shafts should look like this

Reassembly

Refer to illustrations 29.11a, 29.11b and 29.11c

11   Reassembly of the mainshaft is the reverse of disassembly, with the following additions:

a) *Always use new snap-rings. The sharp side of the snap-ring faces away from the thrust washer; the rounded side faces towards the thrust washer.*

b) *With fifth pinion gear, the thrust washer and snap-ring and third-fourth pinion gear on the mainshaft, press sixth and second gears onto the mainshaft together* **(see illustration)**. *Check the side clearance between the gears to be sure they aren't pressed against each other.*

c) *If the ball bearing was pressed off, press it on with the snap-ring groove away from the gears* **(see illustration)**.

d) *Check the positions of the gears on the assembled shaft to make sure they are correct* **(see illustration)**.

Driveshaft

Disassembly

Refer to illustrations 29.12a through 29.12g

12   To disassemble the driveshaft, **refer to illustration 29.3a and the accompanying illustrations**.

29.12a Remove the cup, bearing and thrust washer from the end of the shaft . . .

29.12b . . . slide off first wheel gear . . .

29.12c ... fifth wheel gear ...

29.12d ... remove the snap-ring and thrust washer (arrow) ...

29.12e ... slide off fourth wheel gear ...

29.12f ... third wheel gear ...

Inspection

13   Refer to steps 6 through 10 above to inspect the driveshaft components. If the ball bearing and seal collar need to be replaced, remove them with a press; otherwise, they can be left on the shaft.

Reassembly

14   Assembly is the reverse of the disassembly procedure. Use new snap-rings and lubricate the components with engine oil before assembling them. Check the assembled shaft to make sure the gears are in the correct positions **(see illustration 29.11c)**.

---

30   Shift cam and forks - removal, inspection and installation

Removal

Refer to illustrations 30.3 and 30.4

1   Remove the engine, separate the crankcase halves and remove the transmission shafts (see Sections 5, 19 and 22).
2   Support the shift forks and pull the guide bar out **(see illustration 22.8a)**.
3   Lift out the shift forks and place them in their installed positions on the guide bar so they can be reinstalled correctly **(see illustration)**.
4   Remove the shift cam retainer from the left side of the case **(see illustration)**. Remove the neutral switch from the right side of the case (see Chapter 8).
5   Pull the shift cam out of the case.

Inspection

6   Check the edges of the grooves in the shift cam for signs of excessive wear. Check the pins on each end of the shift cam for wear

29.12g ... and sixth wheel gear - the collar, bearing and second wheel gear will have to be pressed off if they're worn or damaged

and damage. If undesirable conditions are found, replace the shift cam and pins as an assembly.
7   Check the shift forks for distortion and wear, especially at the fork tips. If they are discolored or severely worn they are probably bent. If damage or wear is evident, check the shift fork groove in the corresponding gear as well. Inspect the guide pins and the shaft bore for excessive wear and distortion and replace any defective parts with new ones.
8   Check the shift fork guide bar for evidence of wear, galling and other damage. Make sure the shift forks move smoothly on the bar. If the bar is worn or bent, replace it with a new one.

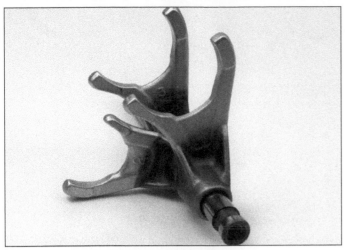

30.3 Reassemble the forks onto the guide bar so they don't get mixed up; the number on each fork faces the right side of the engine when the forks are installed

30.4 Bend back the lockwasher tabs and unscrew the guide bar (A); you'll probably need an impact driver to remove the screw (B)

31.2 Remove the Allen bolts (arrows) and bearing retainer

31.3 Remove the oil nozzle (arrow)

Installation

9 Installation is the reverse of removal, noting the following points:
   a) Lubricate all parts with engine oil before installing them.
   b) The shift forks are numbered 1 through 3, starting from the right side of the engine. The numbers cast in the forks face the right side of the engine.
   c) Engage the follower pin on each shift fork with the shift cam as you pass the guide bar through the fork.
   d) Apply a non-permanent locking agent to the threads of the shift cam retainer screws and tighten them to the torque listed in this chapter's Specifications.

31 Starter clutch and drive chain tensioner - removal, disassembly, inspection and installation

Removal

Refer to illustrations 31.2, 31.3, 31.4 and 31.6

1   Remove the engine, separate the crankcase halves and remove the transmission shafts (see Sections 5, 22 and 28).
2   Refer to Section 22 and remove the primary drive gear, then remove the bearing retainer from behind the gear **(see illustration)**.
3   On the other side of the engine, remove the bearing retainer and oil spray nozzle **(see illustration)**.

31.4 Remove the bearing housing, together with the oil seal

4   Remove the bearing housing from the case. Hold the starter clutch with one hand and pull out the starter clutch shaft **(see illustration)**.
5   Lift the starter idle gear from the case **(see illustration 22.8b)**.

31.6 Lift the starter clutch out of the crankcase

31.10a Push the pins away from the rollers with a small
screwdriver and take the rollers out . . .

31.10b . . . then remove the pins and springs

31.10c The pins should fit against the rollers like this

6   Lift the starter clutch from the drive chain and lift it out **(see illustration)**.
7   Remove the Allen bolts that hold the chain tensioner pad to the case and remove the tensioner **(see illustration 22.8a)**.
8   Remove the Allen bolts that hold the starter chain tensioner **(see illustration 22.8a)** and unscrew the oil pressure relief valve used to secure the oil supply line (see Section 16).

Disassembly, inspection and reassembly

Refer to illustrations 31.10a, 31.10b and 31.10c

9   Refer to **illustration 22.8b** and separate the starter clutch components. Check all parts for wear and damage and replace as necessary.
10   Press a starter clutch pin away from the roller with a small screwdriver, against the spring tension, and remove the roller **(see illustration)**. Remove the pin and spring **(see illustration)**. Remove the other rollers, springs and pins in the same way. Inspect the rollers, springs and pins and replace any that show wear or damage. Reassemble the starter clutch so the pins retain the rollers in position **(see illustration)**.
11   Rotate the bearing on the starter clutch shaft and check for roughness, looseness or noise. If the bearing is bad, have it pressed off the shaft and a new one pressed on by a Yamaha dealer or machine shop.
12   The oil seal in the bearing housing should be replaced if it's worn.

Since the seal is inexpensive and a good deal of work is required to get to it, it's a good idea to replace the seal as a matter of course. Tap the seal out of the housing with a hammer and a bearing driver or socket, then install a new one with the same tools.
13   Pull the piston and spring assembly from the starter drive chain tensioner housing **(see illustration 22.8b)**.
14   Use a pair of pliers to remove the oil pipe from the bottom of the tensioner housing. Be sure to replace the O-ring whenever the oil pipe is removed.

Installation

15   Installation is the reverse of the removal steps. Use non-permanent thread locking agent on the threads of the starter chain guide bolts, starter chain tensioner bolts and the bearing housing screws.

32 Initial start-up after overhaul

1   Make sure the engine oil level is correct, then remove the spark plugs from the engine. Place the engine kill switch in the Off position and unplug the primary (low tension) wires from the coils.
2   Turn on the key switch and crank the engine over with the starter several times to build up oil pressure. Reinstall the spark plugs, connect the wires and turn the switch to On.

3   Make sure there is fuel in the tank, then turn the fuel tap to the On position and operate the choke.
4   Start the engine and allow it to run at a moderately fast idle until it reaches operating temperature. **Warning**: *If the oil level indicator light comes on while the engine is running, stop the engine immediately.*
5   Check carefully for oil leaks and make sure the transmission and controls, especially the brakes, function properly before road testing the machine. Refer to Section 33 for the recommended break-in procedure.
6   Upon completion of the road test, and after the engine has cooled down completely, recheck the valve clearances (see Chapter 1).

33 Recommended break-in procedure

1   Any rebuilt engine needs time to break-in, even if parts have been installed in their original locations. For this reason, treat the machine gently for the first few miles to make sure oil has circulated throughout the engine and any new parts installed have started to seat.
2   Even greater care is necessary if the engine has been rebored or a new crankshaft has been installed. In the case of a rebore, the engine will have to be broken in as if the machine were new. This means greater use of the transmission and a restraining hand on the throttle until at least 500 miles (800 km) have been covered. There's no point in keeping to any set speed limit - the main idea is to keep from lugging (labouring) the engine and to gradually increase performance until the 500 mile (800 km) mark is reached. These recommendations can be lessened to an extent when only a new crankshaft is installed. Experience is the best guide, since It's easy to tell when an engine is running freely. The following recommendations, which Yamaha provides for new motorcycles, can be used as a guide:

a) *0 to 90 miles (0 to 150 km): Keep engine speed below 5,000 rpm. Turn off the engine after each hour of operation and let it cool for 5 to 10 minutes. Vary the engine speed and Don't use full throttle.*
b) *90 to 300 miles (150 to 500 km): Don't run the engine for long periods above 6,500 rpm. Rev the engine freely through the gears, but Don't use full throttle.*
c) *300 to 600 miles (500 to 1000 km): Don't use full throttle for prolonged periods and Don't cruise at speeds above 8,000 rpm.*
d) *After 600 miles (1,000 km): Full throttle can be used. Don't exceed maximum recommended engine speed (rodline).*

3   If a lubrication failure is suspected, stop the engine immediately and try to find the cause. If an engine is run without oil, even for a short period of time, severe damage will occur.

Notes

Chapter 3
Fuel and exhaust systems

Contents

Specifications

Fuel grade
Unleaded or leaded (according to local regulations), minimum 91 octane (Research method)

Fuel tank capacity
FJ600
  Except California .... 19 liters (5.02 US gal, 4.18 Imperial gal)
  California .... 18.5 liters (4.89 US gal, 4.07 Imperial gal)
FZ600 .... 16 liters (4.2 US gal, 3.5 Imperial gal)
XJ600 .... 19 liters (5.02 US gal, 4.18 Imperial gal)
YX600 Radian .... 12 liters (3.2 US gal (2.6 Imperial gal)

Fuel tank reserve capacity
All except FZ600 .... 2.5 liters (0.66 US gal, 0.55 Imperial gal)
FZ600 .... 3.0 liters (0.79 US gal, 0.66 Imperial gal)

Carburetor type
FJ600, XJ600 .... Mikuni BS32 (four)
FZ600, YX600 Radian .... Mikuni BS30 (four)

Jet sizes

Main jet
FJ600 .... #1, #2 : 105   #3, #4 : 102.5
FZ600 .... 107.5
XJ600 .... #1, #2 : 105   #3, #4 : 102.5
YX600 Radian... 97.5

Main air jet
FJ600 70
FZ600 140
XJ600 70
YX600 Radian... 140

Jet needle (clip position)
FJ600 .... #1, #3, #4 : 4CP4   #2 : 4CP6
FZ600
      US models 4CHP2
      UK models 4CHP3
XJ600 .... #1, #3, #4 : 4CP3-3   #2, #3: 4CP7-3
YX600 Radian... #1, #4 : 4CHP2   #2, #3 : 4CHP4

Needle jet
FJ600 .... N-8
FZ600 .... O-6
XJ600 .... N-8
YX600 Radian... O-6

Pilot jet
FJ600 35
FZ600 30
XJ600 40
YX600 Radian... 30

Pilot screw
FJ600 .... Preset
FZ600
      US models .... Preset
      UK models .... 2-1/2 turns out
XJ600 .... 2 turns out
YX600 Radian... Preset

Valve seat size
FJ600 .... 2.0
FZ600 .... 2.3
XJ600 .... 2.0
YX600 Radian... 2.3

Starter jet
FJ600 .... 42.5
FZ600 .... 22.5
XJ600 .... 42.5
YX600 Radian... 25.0

Carburetor adjustments

Float height
FJ600 .... not specified
FZ600 .... 20.0 +/- 1.0 mm (0.787 +/- 0.020 inch)
XJ600 .... 16.5 - 18.5 mm (0.650 - 0.730 inch)
YX600 Radian... 20.0 +/- 1.0 mm (0.787 +/- 0.020 inch)

Fuel level
FJ600 .... 3.0 mm +/- 1.0 mm (0.12 +/- 0.040) below mixing chamber body line
FZ600 .... 2.0 +/- 0.5 mm (0.080 +/- 0.020 inch) below float chamber line
XJ600
      1984 and 1985 models .... 3.0 +/- 1.0 mm (0.12 +/- 0.040) below mixing chamber body line
      1989 and later models .... 2.5 to 3.5 mm (0.10 to 0.14 inch) below float chamber line
YX600 Radian... 2.0 +/- 0.5 mm (0.08 +/- 0.02 inch) below mixing chamber body line

Tightening torques

Exhaust pipe flange nuts .... 10 Nm (7 ft-lbs)*
Muffler chamber mounting bolt .... 25 Nm (18 ft-lbs)*
Muffler mounting bracket bolts .... 25 Nm (18 ft-lbs)
Exhaust pipe and muffler clamp bolts... 20 Nm (14 ft-lbs)*

*Lubricate the threads with anti-seize compound.

2.9 Remove the mounting bolt at the rear of the tank

2.11a Inspect the mounts at the rear of the tank . . .

1 General information

The fuel system consists of the fuel tank, the fuel tap and filter, the carburetors and the connecting lines, hoses and control cable(s).

The carburetors used on these motorcycles are four BS30 or BS32 Mikunis with butterfly-type throttle valves. For cold starting, an enrichment circuit is provided. The enrichment circuit is actuated by a cable and the choke knob mounted on the left side handlebar (FJ600 and Radian models), a knob attached directly to the carburetors (XJ600 models) or a lever attached directly to the carburetors (FZ600 models).

The exhaust system is either a four-into-two or a four-into-one design.

Many of the fuel system service procedures are considered routine maintenance items and for that reason are included in Chapter 1.

2 Fuel tank - removal and installation

Refer to illustrations 2.9, 2.11a and 2.11b
Warning: *Gasoline (petrol) is extremely flammable, so take extra precautions when you work on any part of the fuel system. don't smoke or allow open flames or bare light bulbs near the work area, and don't work in a garage where a natural gas-type appliance (such as a water heater or clothes dryer) is present. If you spill any fuel on your skin, rinse it off immediately with soap and water. When you perform any kind of work on the fuel system, wear safety glasses and have a fire extinguisher suitable for a class B type fire (flammable liquids) on hand.*
1   Remove the seat (see Chapter 7).
2   Disconnect the cable from the negative terminal of the battery.
3   If you're working on an XJ600 or FJ600, remove the side covers (see Chapter 7).
4   If you're working on an FZ600, remove the side covers and center and lower fairings (see Chapter 7).
5   If you're working on a YX600 Radian, remove the fuel tap cover.
6   Detach the fuel line and vacuum line from the tank.
7   On all except FZ600 models, detach the breather hose. On FZ600 models, disconnect the fuel gauge electrical connector.
8   The fuel tank is held in place at the forward end by two cups, one on each side of the tank, which slide over two rubber dampers on the frame. The rear of the tank is fastened to a bracket by one bolt and a rubber insulator, which fit through a flange projecting from the tank.
9   Remove the tank mounting bolt **(see illustration)**.
10 Lift the rear of the tank up. If you're working on an FZ600, disconnect the breather hose. Slide the tank to the rear to disengage the front of the tank from the rubber dampers, then carefully lift the tank away from the machine.
11 Before installing the tank, check the condition of the rubber mounting dampers **(see illustrations)** - if they're hardened, cracked,

2.11b . . . and at the front

or show any other signs of deterioration, replace them.
12 When replacing the tank, reverse the above procedure. Make sure the tank seats properly and does not pinch any control cables or wires. If difficulty is encountered when trying to slide the tank cups onto the dampers, a small amount of light oil should be used to lubricate them.

3 Fuel tank - cleaning and repair

1   All repairs to the fuel tank should be carried out by a professional who has experience in this critical and potentially dangerous work. Even after cleaning and flushing of the fuel system, explosive fumes can remain and ignite during repair of the tank.
2   If the fuel tank is removed from the machine, it should not be placed in an area where sparks or open flames could ignite the fumes coming out of the tank. Be especially careful inside garages where a natural gas-type appliance is located, because the pilot light could cause an explosion.

4 Idle fuel/air mixture adjustment - general information

1   Due to the increased emphasis on controlling motorcycle exhaust emissions, certain governmental regulations have been formulated which directly affect the carburetion of this machine. In order to comply with the regulations, the carburetors on some models have a metal sealing plug pressed into the hole over the pilot screw (which controls the idle fuel/air mixture) on each carburetor, so they can't be tampered with. These should only be removed in the event of a

complete carburetor overhaul, and even then the screws should be returned to their original settings. The pilot screws on other models are accessible, but the use of an exhaust gas analyzer is the only accurate way to adjust the idle fuel/air mixture and be sure the machine doesn't exceed the emissions regulations.

2   If the engine runs extremely rough at idle or continually stalls, and if a carburetor overhaul does not cure the problem, take the motorcycle to a Yamaha dealer service department or other repair shop equipped with an exhaust gas analyzer. They will be able to properly adjust the idle fuel/air mixture to achieve a smooth idle and restore low speed performance.

5   Carburetor overhaul - general information

1   Poor engine performance, hesitation, hard starting, stalling, flooding and backfiring are all signs that major carburetor maintenance may be required.

2   Keep in mind that many so-called carburetor problems are really not carburetor problems at all, but mechanical problems within the engine or ignition system malfunctions. Try to establish for certain that the carburetors are in need of maintenance before beginning a major overhaul.

3   Check the fuel filter, the fuel lines, the tank cap vent (except California models), the intake manifold hose clamps, the vacuum hoses, the air filter element, the cylinder compression, the spark plugs, and the carburetor synchronization before assuming that a carburetor overhaul is required.

4   Most carburetor problems are caused by dirt particles, varnish and other deposits which build up in and block the fuel and air passages. Also, in time, gaskets and O-rings shrink or deteriorate and cause fuel and air leaks which lead to poor performance.

5   When the carburetor is overhauled, it is generally disassembled completely and the parts are cleaned thoroughly with a carburetor cleaning solvent and dried with filtered, unlubricated compressed air. The fuel and air passages are also blown through with compressed air to force out any dirt that may have been loosened but not removed by the solvent. Once the cleaning process is complete, the carburetor is reassembled using new gaskets, O-rings and, generally, a new inlet needle valve and seat.

6   Before disassembling the carburetors, make sure you have a carburetor rebuild kit (which will include all necessary O-rings and other parts), some carburetor cleaner, a supply of rags, some means of blowing out the carburetor passages and a clean place to work. It is recommended that only one carburetor be overhauled at a time to avoid mixing up parts.

6   Carburetors - removal and installation

Warning: *Gasoline (petrol) is extremely flammable, so take extra precautions when you work on any part of the fuel system. don't smoke or allow open flames or bare light bulbs near the work area, and don't work in a garage where a natural gas-type appliance (such as a water heater or clothes dryer) is present. If you spill any fuel on your skin, rinse it off immediately with soap and water. When you perform any kind of work on the fuel system, wear safety glasses and have a fire extinguisher suitable for a class B type fire (flammable liquids) on hand.*

Removal

Refer to illustrations 6.4 and 6.23

1   Remove the fuel tank (see Section 2).

2   Disconnect the negative cable from the battery.

XJ/FJ600 models

3   Disconnect the throttle cable (and choke cable if equipped) (see Sections 10 and 11).

4   Loosen the clamping bands on the air cleaner joints and intake joints (the rubber tubes that secure the carburetors to the air cleaner air box and engine **(see illustration)**.

6.4 Loosen the clamping bands around the air box joints (arrow) and the intake manifold tubes

5   Remove the air cleaner case bolts and slide the air cleaner case backward to disengage the air cleaner joints from the carburetors.

6   Slide the carburetors backward to disengage them from the intake joints, then remove the carburetor assembly.

FZ600 models

7   Remove the starter relay, the regulator/rectifier and the battery (see Chapter 8).

8   Loosen the clamping bands on the air cleaner joints and intake joints (the rubber tubes that secure the carburetors to the air cleaner air box and engine **(see illustration 6.4)**.

9   Remove the air cleaner case bolts and slide the air cleaner case backward to disengage the air cleaner joints from the carburetors.

10   Disconnect the throttle cable from the throttle pulley (see Section 10).

11   Disconnect the air vent hose from the carburetor assembly.

12   Remove the fairing stay.

13   Slide the carburetors backward to disengage them from the intake joints, then remove the carburetor assembly.

YX600 Radian models

14   If you're working on a YX600 Radian, remove the carburetor covers and side covers (see Chapter 7).

15   Remove the fuse block mounting screws (see Chapter 7) and remove the igniter unit (see Chapter 4).

16   Remove the battery case bolts and air cleaner air box bolts.

17   Loosen the clamping bands on the air cleaner joints and intake joints (the rubber tubes that secure the carburetors to the air cleaner air box and engine **(see illustration 6.4)**.

18   Slide the air cleaner air box and the battery case backward to disengage the air cleaner joints from the carburetors.

19   Disconnect the choke cable and throttle cable from the carburetor assembly (see Sections 10 and 11). Disconnect the air vent hose.

20   Loosen the clamp screws on the intake manifolds (the rubber tubes that connect the carburetors to the engine) **(see illustration 6.4)**.

21   Slide the carburetors backward to disengage them from the intake joints, then remove them.

All models

22   After the carburetors have been removed, stuff clean rags into the intake manifold tubes to prevent the entry of dirt or other objects.

23   Inspect the rubber intake joints **(see illustration)**. If they're cracked or brittle, replace them.

Installation

24   Installation is the reverse of the removal Steps, with the following additions:
  a) *Adjust the throttle grip freeplay (see Chapter 1).*
  b) *Check and, if necessary, adjust the idle speed and carburetor synchronization (see Chapter 1).*

6.23 Inspect the intake joints and replace them if they're cracked or brittle

7   Carburetors - disassembly, cleaning and inspection

Warning: *Gasoline (petrol) is extremely flammable, so take extra precautions when you work on any part of the fuel system. Don't smoke or allow open flames or bare light bulbs near the work area, and Don't work in a garage where a natural gas-type appliance (such as a water heater or clothes dryer) is present. If you spill any fuel on your skin, rinse it off immediately with soap and water. When you perform any kind of work on the fuel system, wear safety glasses and have a fire extinguisher suitable for a class B type fire (flammable liquids) on hand.*

Disassembly

Refer to illustrations 7.1a, 7.1b, 7.2, 7.3a, 7.3b, 7.4, 7.5, 7.6a, 7.6b, 7.6c, 7.6d, 7.7, 7.8, 7.9, 7.10a, 7.10b, 7.11a, 7.11b, 7.12a, 7.12b, 7.13a, 7.13b, 7.14a, 7.14b, 7.15a, 7.15b, 7.16, 7.17a, 7.17b, 7.17c, 7.18a through 7.18e, 7.21 and 7.22

1   Remove the carburetors from the machine as described in Section 6. Set the assembly on a clean working surface. **Note**: *Unless the O-rings on the fuel and vent fittings between the carburetors are leaking, Don't detach the carburetors from their mounting brackets. Also, work on one carburetor at a time to avoid getting parts mixed up. it's possible to overhaul the carburetors without removing them from the brackets* **(see illustrations)**.

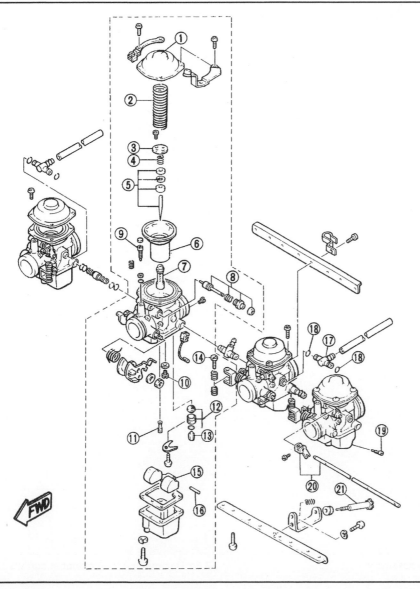

7.1a Carburetors (FJ600 and XJ600 models) - exploded view

1	Vacuum chamber cover
2	Throttle valve spring
3	Jet needle holder
4	Jet needle spring
5	Jet needle set
6	Throttle valve
7	Main nozzle
8	Choke plunger
9	Pilot screw
10	Main jet
11	Pilot jet
12	Needle valve seat
13	Needle valve
14	Synchronizing screw
15	Float
16	Float pivot pin
17	Vent hose joint
18	O-ring
19	Float chamber drain screw
20	Choke lever
21	Throttle stop screw

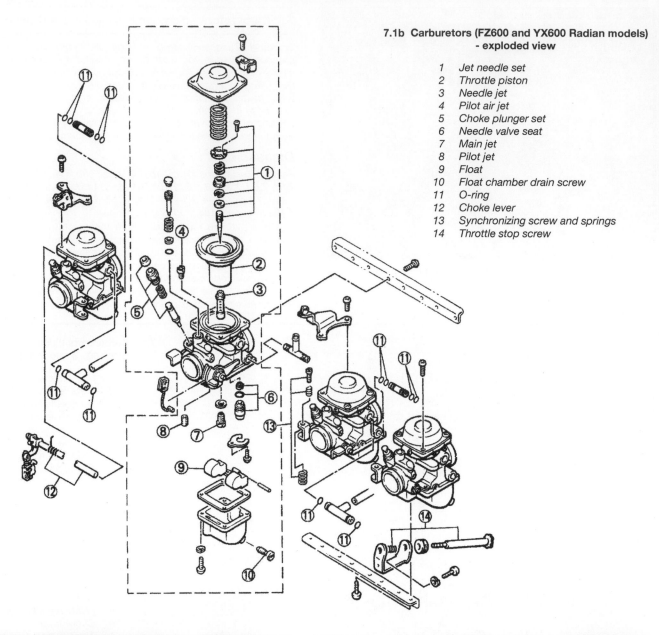

7.1b Carburetors (FZ600 and YX600 Radian models) - exploded view

1 Jet needle set
2 Throttle piston
3 Needle jet
4 Pilot air jet
5 Choke plunger set
6 Needle valve seat
7 Main jet
8 Pilot jet
9 Float
10 Float chamber drain screw
11 O-ring
12 Choke lever
13 Synchronizing screw and springs
14 Throttle stop screw

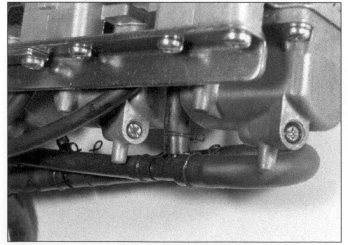

7.2 Remove the vent hose assembly

7.3a Remove the float chamber screws . . .

7.3b . . . and lift off the float chamber

7.4 Push out the float pivot pin

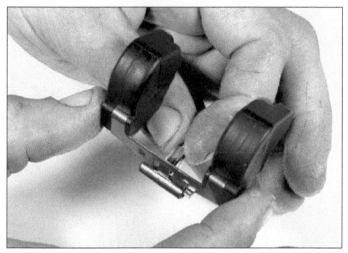

7.5 Lift out the float and unhook the needle valve

7.6a Remove the screw and retainer . . .

2   Remove the vent hose assembly (see illustration).
3   Remove the float chamber screws and lift off the float chamber (see illustrations).
4   Push the float pivot pin out with a thin punch (see illustration).
5   Lift out the float and unhook the needle valve (see illustration).

6   Remove the retainer, then pull out the needle valve seat and remove the filter screen (see illustrations). Remove any dirt build-up on the filter screen.
7   Unscrew the main jet (see illustration). don't lose the washer.
8   Unscrew the pilot jet (see illustration).

7.6b . . . and pull out the needle valve seat

7.6c This is the typical type of debris found on the needle and seat filter screen

7.6d Remove the filter screen

7.7 Unscrew the main jet and lift it out together with its washer

7.8 Remove the pilot jet

7.9 Remove all traces of the old float chamber gasket

7.10a Remove the vacuum chamber cover screws . . .

9   Remove the float chamber gasket **(see illustration)**.
10   Remove the screws and lift off the vacuum chamber cover **(see illustrations)**.

11   Lift out the spring, then carefully separate the diaphragm from the carburetor without tearing it and lift out the throttle piston **(see illustrations)**. Note the position of the rubber diaphragm.

7.10b . . . and lift off the cover

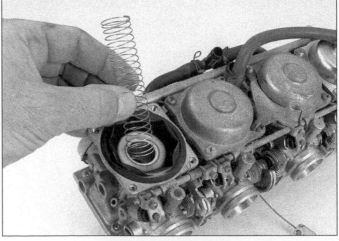

7.11a Lift out the spring

7.11b Peel the diaphragm away slowly so you Don't tear it, then remove the diaphragm and throttle piston; align the diaphragm tab with the notch in the carburetor body (arrow) on assembly

7.12a Remove the two screws inside the throttle piston . . .

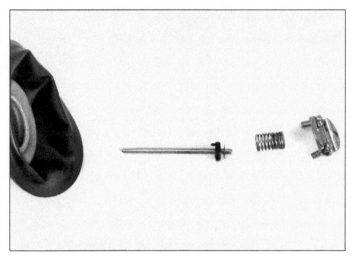

7.12b . . . and lift out the plate that secures the jet needle, then lift out the spring and jet needle

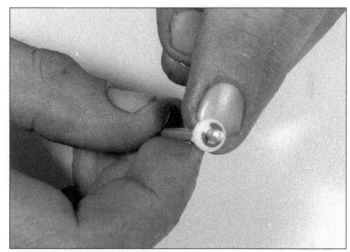

7.13a Remove the clip from the jet needle . . .

12 Remove two screws from inside the throttle piston and take out the plate **(see illustration)**. Remove the spring and jet needle **(see illustration)**.
13 Remove the clip and washer from the jet needle (if there's more than one groove on the needle, note which one the clip was fitted to) **(see illustrations)**.
14 Push the needle jet into the throttle bore with a punch or similar tool, then take it out **(see illustrations)**.

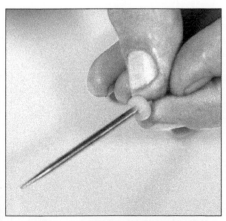

7.13b . . . and take off the washer

7.14a Push the needle jet into the throttle bore . . .

7.14b . . . and take it out through the throttle opening

7.15a Loosen the pilot air jet . . .

7.15b . . . and lift it out of the carburetor housing

7.16 Loosen the choke rod set screws and slide out the choke rod (FZ600 and YX600 Radian shown; XJ600 and FJ600 similar)

7.17a Unscrew the choke plunger cap with a box wrench (ring spanner) (FZ600 and YX600 Radian shown; XJ600 and FJ600 similar) . . .

15 Remove the pilot air jet from the carburetor body **(see illustrations)**.
16 Loosen the choke shaft screws and slide the choke shaft out of the brackets **(see illustration)**. Note the detents in the rod used to

position the screws in the correct location.
17 Unscrew the choke plungers and pull them out of their bores **(see illustrations)**.

7.17b . . . and pull out the spring . . .

7.17c . . . and the choke plunger

7.18a Drill out the pilot screw plug, but be very careful not to drill into the screw

7.18b The easiest way to remove the plug is with a small slide hammer, but you can pry it out with a screwdriver

7.18c Turn the screw (arrow) clockwise until it bottoms lightly - count the number of turns and write it down . . .

18 The pilot (idle mixture) screw is located in the front of the carburetor body. On US models, this screw is hidden behind a plug which will have to be removed if the screw is to be taken out. To do this, drill a hole in the plug, being careful not to drill into the screw, then pry the plug out or remove it with a small slide hammer (see illustrations).

On all models, turn the pilot screw in, counting the number of turns until it bottoms lightly (see illustration). Record that number for use when installing the screw. Now remove the pilot screw along with its spring, washer and O-ring (see illustration).

Cleaning

Caution: *Use only a petroleum based solvent for carburetor cleaning. Never use caustic cleaners. Be sure that all rubber hoses and plastic fittings are removed before the parts are immersed into the cleaning agent.*

19 Submerge the metal components in the solvent for approximately thirty minutes (or longer, if the directions recommend it).

20 After the carburetor has soaked long enough for the cleaner to loosen and dissolve most of the varnish and other deposits, use a brush to remove the stubborn deposits. Rinse it again, then dry it with compressed air. Blow out all of the fuel and air passages in the main and upper body. **Caution**: *Never clean the jets or passages with a piece of wire or a drill bit, as they will be enlarged, causing the fuel and air metering rates to be upset.*

Inspection

Refer to illustrations 7.21 and 7.22

21 Check the operation of the choke plunger. If it doesn't move smoothly, replace it, along with the return spring. Inspect the needle on the end of the choke plunger and replace it if it's worn (see illustration).

7.18d . . . then remove the screw . . .

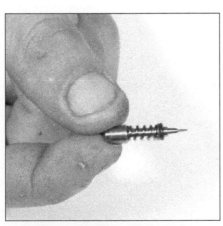

7.18e . . . together with the spring, washer and O-ring

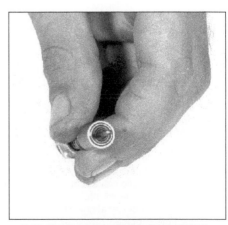

7.21 Replace the choke plunger if It's worn at the tip

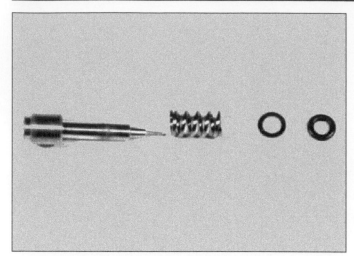

7.22 Replace the pilot screw if It's worn or scratched at the tip; replace the spring if It's weak and replace the washers if they're damaged

8.3a Note the position of the springs on the synchronizer screws - they should look like this when installed - this screw is between the no. 3 and no. 4 carburetors . . .

8.3b . . . this screw is between the center carburetors (no. 2 and no. 3) . . .

22   Check the tapered portion of the pilot screw for wear or damage **(see illustration)**. Replace the pilot screw if necessary.
23   Check the carburetor body, float chamber and vacuum chamber cover for cracks, distorted sealing surfaces and other damage. If any defects are found, replace the faulty component, although replacement of the entire carburetor will probably be necessary (check with your parts supplier for the availability of separate components).
24   Check the diaphragm for splits, holes and general deterioration. Holding it up to a light will help to reveal problems of this nature.
25   Insert the throttle piston in the carburetor body and see that it moves up-and-down smoothly. Check the surface of the valve for wear. If it's worn excessively or doesn't move smoothly in the bore, replace the carburetor.
26   Check the jet needle for straightness by rolling it on a flat surface (such as a piece of glass). Replace it if it's bent or if the tip is worn.
27   Check the tip of the fuel inlet valve needle. If it has grooves or scratches in it, it must be replaced. Push in on the rod in the other end of the needle, then release it - if it doesn't spring back, replace the valve needle. Inspect the filter screen on the needle valve seat and replace it if it's torn or can't be cleaned.
28   Check the gasket on the float bowl. Replace it if it's damaged.
29   Operate the throttle shaft to make sure the throttle butterfly valve opens and closes smoothly. If it doesn't, replace the carburetor.
30   Check the floats for damage. This will usually be apparent by the presence of fuel inside one of the floats. If the floats are damaged, they must be replaced.

Reassembly

Caution: *When installing the jets, be careful not to over-tighten them - they're made of soft material and can strip or shear easily.*
Note: *When reassembling the carburetors, be sure to use the new O-rings, gaskets and other parts supplied in the rebuild kit.*
31   Install the choke plunger in its bore, followed by its spring and cap. Tighten the cap securely.
32   Install the pilot screw (if removed) along with its spring, washer and O-ring, turning it in until it seats lightly. Now, turn the screw out the number of turns that was previously recorded (or the number listed in this Chapter's Specifications where applicable). On US models install a new metal plug in the hole over the screw. Apply a little bonding agent around the circumference of the plug after it has been seated.
33   Install the needle valve seat, retainer and screw.
34   Install the needle jet into its hole in the bottom of the carburetor. The slot in the needle jet aligns with a pin in the carburetor body.
35   Install the pilot jet.
36   Install the main jet with its washer. Hook the needle valve over the float, then install the float and secure it with the pivot pin.

37   Refer to Section 9 and check the float height.
38   Install the washer and clip on the jet needle, then install the jet needle and plate in the throttle piston. If there's more than one clip groove on the jet needle, install the clip in the groove listed in this Chapter's Specifications. Attach the plate with the two screws (a magnetic screwdriver will make this easier).
39   Install the diaphragm and piston in the carburetor body. Be sure the protrusion on the edge of the diaphragm fits into its groove all the way around.
40   Install the vacuum chamber cover and tighten its screws securely.
41   Slide the choke shaft into its brackets. Align the drilled notches in the shaft with the screws, then tighten the screws securely.

8   Carburetors - separation and joining

Refer to illustrations 8.3a, 8.3b and 8.3c

1   The carburetors do not need to be separated for normal overhaul. If you need to separate them (to replace a carburetor body, for example), refer to the following procedure.
2   Remove the bracket screws **(see illustration 7.1a or 7.1b)**. You may need an impact driver.

8.3c . . . and this screw is between the no. 1 and no. 2 carburetors

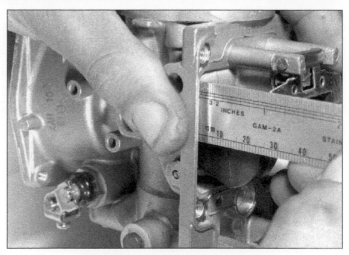

9.1 Measure float height with a ruler

3   Note how the synchronizing screws and springs are assembled **(see illustrations)**. As you pull the carburetors apart, keep track of the synchronizer screw springs. They should stay with the adjusting screws, but if they don't, find them and install them as shown in the illustrations so they aren't lost.

4   Assembly is the reverse of the disassembly procedure. Use non-permanent thread locking agent on the bracket screws.

9   Carburetors - float height and fuel level adjustment

Float height adjustment

Refer to illustration 9.1

1   To check the float height, hold the carburetor so the float hangs down, then tilt it back until the valve needle is just seated (the small rod in the end of the needle valve shouldn't be compressed). Measure the distance from the carburetor body (without the float chamber gasket) to the top of the float **(see illustration)** and compare your measurement to the float height listed in this Chapter's Specifications. If it isn't as specified, carefully bend the tang that contacts the valve needle up or down until the float height is correct.

Fuel level adjustment

Refer to illustration 9.4

Warning: *Gasoline (petrol) is extremely flammable, so take extra precautions when you work on any part of the fuel system. don't smoke or allow open flames or bare light bulbs near the work area, and don't work in a garage where a natural gas-type appliance (such as a water heater or clothes dryer) is present. If you spill any fuel on your skin, rinse it off immediately with soap and water. When you perform any kind of work on the fuel system, wear safety glasses and have a fire extinguisher suitable for a class B type fire (flammable liquids) on hand.*

2   Install the carburetor assembly (see Section 6) and check the fuel level.

3   Place the bike on its centerstand (if equipped) or support it securely upright. Place a floor jack under the engine and raise it just enough so the carburetors are horizontal.

4   Attach Yamaha service tool no. YM-01312 (US) or 90890-01312 (UK) to the drain fitting on the bottom of one of the carburetor float chambers (all four will be checked). This is a clear plastic tube graduated in millimeters **(see illustration)**. An alternative is to use a length of clear plastic tubing and an accurate ruler. Hold the graduated tube (or the free end of the clear plastic tube) against the carburetor body.

5   Unscrew the drain screw at the bottom of the float chambers a couple of turns, then start the engine and let it idle - fuel will flow into

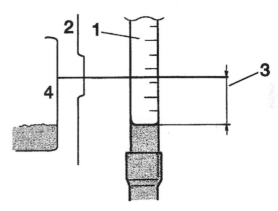

9.4 A gauge like this or a clear tube and ruler can be used to measure fuel level

1	Gauge	3	Fuel level
2	Carburetor body	4	Float chamber

the tube. Wait for the fuel level to stabilize, then shut the engine off and note how far the fuel level is below the line where the float chamber joins the carburetor body.

6   Measure the distance between the joining line of the float chamber/carburetor body and the top of the fuel level in the tube or gauge. This distance is the fuel level - write it down on a piece of paper, screw in the drain screw, then move on to the next carburetor and check it the same way.

7   Compare your fuel level readings to the value listed in this Chapter's Specifications. If the fuel level in any carburetor is not correct, remove the float chamber and bend the float tang up or down as necessary, then recheck the fuel level.

10   Throttle cable and grip - removal, installation and adjustment

Refer to illustrations 10.3 and 10.5

Removal

1   Remove the fuel tank (see Section 2).

2   Loosen the accelerator cable (see *Throttle operation/grip freeplay - check and adjustment* in Chapter 1).

3   Remove the screws from the throttle switch housing and separate the halves of the housing **(see illustration)**.
4   Detach the accelerator cable from the throttle grip pulley. Unscrew the grip end weight (if equipped) and take the throttle grip off the handlebar.
5   Detach the other end of the cable from the throttle pulley at the carburetors **(see illustration)**.
6   Remove the cable, noting how it is routed.

Installation

7   Clean the handlebar and apply a light coat of multi-purpose grease.
8   Push the grip onto the handlebar. Install the grip end weight (if equipped).
9   Route the cable into place. Make sure doesn't interfere with any other components and isn't kinked or bent sharply.
10   Lubricate the ends of the accelerator cable with multi-purpose grease and connect it to the throttle pulleys at the carburetors and at the throttle grip.

Adjustment

11   Follow the procedure outlined in Chapter 1, Throttle operation/ grip freeplay - check and adjustment, to adjust the cables.
12   Turn the handlebars back and forth to make sure the cable doesn't cause the steering to bind. With the engine idling, turn the handlebars back and forth and make sure idle speed doesn't change. If it does, find and fix the cause before riding the motorcycle.
13   Install the fuel tank.

---

11   Choke cable - removal and installation

A choke cable is used on FJ600 and YX600 Radian models.

Removal

Refer to illustrations 11.2 and 11.4

FJ600 models

1   If you're working on an FJ600, remove the choke lever screw and detach the choke lever from the left handlebar grip.

YX600 Radian models

2   Remove the choke thumb lever screw with a small Phillips screwdriver **(see illustration)**. **Note:** *This screw has been secured with thread locking agent, so it's very tight. Be careful not to strip it out. Once you've removed the screw, thread it back into its hole temporarily so it doesn't get lost.*

10.3 Remove the screws (arrows) and separate the halves of the throttle switch housing (upper screw obscured by cable housing)

All models

3   If necessary for access, remove the fuel tank (see Section 2).
4   Loosen the Phillips screw that holds the choke cable housing on the carburetor **(see illustration)**.
5   Slip the lug end from the choke lever.

Installation

6   Installation is the reverse of the removal steps. Apply a non-permanent thread locking agent to the threads of the choke knob screw.

---

12   Exhaust system - removal and installation

Refer to illustrations 12.2, 12.3, 12.4, 12.8 and 12.9

1   There are three different types of exhaust systems used on these models.
2   The FJ600 and XJ600 use a four-into-two exhaust system with crossover pipes **(see illustration)**.
3   The FZ600 uses a four-into-one exhaust system **(see illustration)**.
4   The YX600 Radian uses a four-into-two exhaust system with a center chamber **(see illustration)**.
5   Set the machine on the center stand (if equipped) or prop it securely upright.

10.5 Pull up the throttle pulley to create slack in the cable, then align the cable with the slot in the pulley and slip it out

11.2 The screw that retains the choke lever on YX600 Radian models has been secured with thread locking agent; be sure the screwdriver fits correctly

11.4 Loosen the screw (arrow) to remove the choke cable from the bracket, then turn the cable sideways and slip it out of the lever

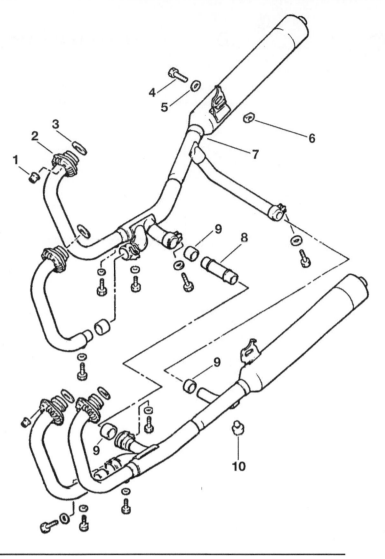

12.2 Exhaust system (FJ600 and XJ600 models) - exploded view

1   Retaining nut
2   Retaining ring
3   Gasket
4   Mounting bolt
5   Washer
6   Nut
7   Exhaust pipe and
    muffler/silencer assembly
8   Crossover pipe
9   Gasket
10 Centerstand stopper

12.3 Exhaust system (FZ600 models)- exploded view

1   Retaining nut
2   Retaining ring
3   Gasket
4   Exhaust pipe and
    muffler/silencer assembly
5   Bolt
6   Washer
7   Grommet
8   Washer
9   Nut
10 Sealing washer
11 Plug

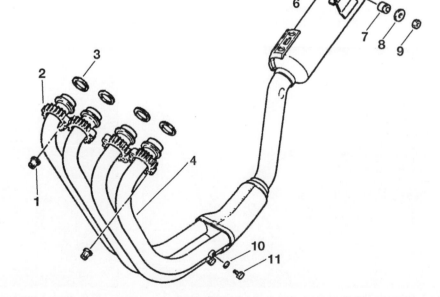

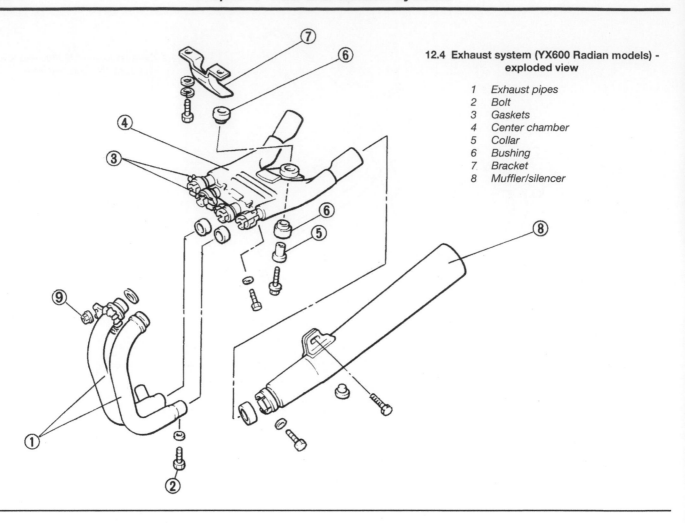

12.4 Exhaust system (YX600 Radian models) - exploded view

1   *Exhaust pipes*
2   *Bolt*
3   *Gaskets*
4   *Center chamber*
5   *Collar*
6   *Bushing*
7   *Bracket*
8   *Muffler/silencer*

6   If you're working on an FJ600 or XJ600, remove the lower fairing (see Chapter 7). If you're working on an FZ600, remove the center and lower fairings.
7   Support the exhaust system with a jack.
8   Remove the exhaust pipe holder nuts at the cylinder head and slide the holders off the mounting studs **(see illustration)**.
9   Remove the exhaust system mounting fasteners. Pull the exhaust pipes forward to clear the engine **(see illustration)**, then lower the jack and pull the exhaust system out from under the motorcycle.

10   If necessary, loosen the clamp bolts and separate the exhaust system components.
11   Installation is the reverse of removal, with the following additions:
 a) *Use new gaskets at the cylinder head* **(see illustration 12.9)**.
 b) *Apply a thin coat of anti-seize compound to the studs and nuts before installation.*
 c) *Tighten all fasteners to the torque settings listed in this Chapter's Specifications.*

12.8 Remove the mounting nuts at the cylinder head

12.9 Always replace the copper gaskets when reconnecting the exhaust pipes to the cylinder head

Chapter 4
Ignition system

Contents

Specifications

Ignition coil

Ignition coil primary resistance (at 20-degrees C/68-degrees F)	
All except 1989 and later XJ600	2.43 to 2.97 ohms
1989 and later XJ600	1.8 to 2.2 ohms
Ignition coil secondary resistance (at 20-degrees C/68-degrees F)	
FJ600, XJ600	9600 to 14,400 ohms
FZ600, YX600 Radian	10,560 to 15,840 ohms

Spark plug caps and spark plugs

Spark plug cap resistance	10,000 ohms
Spark plug arcing distance	6 mm (1/4 inch)

Pickup coil resistance (at 20-degrees C (68-degrees F)

All except 1989 and later XJ600 (black to gray and black to orange)	108 to 132 ohms
1989 and later XJ600 (white/red to white/black)	81 to 121 ohms

Ignition timing

Electronic, not adjustable

Firing order

1 - 2 - 4 - 3

Tightening torques

Early models	
Pickup coil plate securing screws	8 Nm (5.8 ft-lbs)*
Timing rotor bolt	24 Nm (17 ft-lbs)
Signal generator cover screws	10 Nm (7.2 ft-lbs)
Later models	
Alternator cover bolts	See Chapter 8
Stator and pickup coil screws	See Chapter 8

*Apply gasket sealant to the screw threads.

2.14 A simple spark gap testing fixture can be made from a block of wood, a large alligator clip, two nails, a screw and a piece of wire

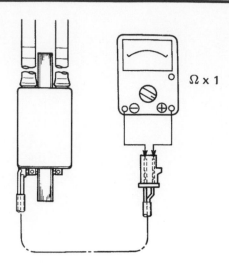

$\Omega \times 1$

3.4 To test the coil primary resistance, connect the ohmmeter leads between the primary terminals in the coil connector

1   General information

This motorcycle is equipped with a battery operated, fully transistorized, breakerless ignition system. The system consists of the following components:

Pickup coils (early models) or single pickup coil (later models)
Igniter unit
Battery and fuse
Ignition coils
Spark plugs
Ignition (main) and engine kill (stop) switches
Primary and secondary circuit wiring

The transistorized ignition system functions on the same principle as a breaker point DC ignition system with the pickup unit and igniter performing the tasks previously associated with the breaker points and mechanical advance system. As a result, adjustment and maintenance of ignition components is eliminated (with the exception of spark plug replacement).

Early models use two pickup coils to control ignition timing. Later models use a digital microprocessor and a single pickup coil.

Because of their nature, the individual ignition system components can be checked but not repaired. If ignition system troubles occur, and the faulty component can be isolated, the only cure for the problem is to replace the part with a new one. Keep in mind that most electrical parts, once purchased, can't be returned. To avoid unnecessary expense, make very sure the faulty component has been positively identified before buying a replacement part.

2   Ignition system - check

Refer to illustration 2.14
Warning: *Because of the very high voltage generated by the ignition system, extreme care should be taken when these checks are performed.*
1   If the ignition system is the suspected cause of poor engine performance or failure to start, a number of checks can be made to isolate the problem.
2   Make sure the engine kill switch is in the Run position.

Engine will not start

3   Disconnect one of the spark plug wires, connect the wire to a spare spark plug and lay the plug on the engine with the threads contacting the engine. If necessary, hold the spark plug with an insulated tool. Crank the engine over and make sure a well-defined, blue spark occurs between the spark plug electrodes.

Warning: *Don't remove one of the spark plugs from the engine to perform this check - atomized fuel being pumped out of the open spark plug hole could ignite, causing severe injury!*
4   If no spark occurs, the following checks should be made:
5   Unscrew a spark plug cap from a plug wire. Check the cap resistance with an ohmmeter and compare it to the value listed in this Chapter's Specifications. If the resistance is infinite, replace it with a new one. Repeat this check on the remaining plug caps.
6   Make sure all electrical connectors are clean and tight. Check all wires for shorts, opens and correct installation.
7   Check the battery voltage with a voltmeter and check the specific gravity with a hydrometer (see Chapter 1). If the voltage is less than 12-volts or if the specific gravity is low, recharge the battery.
8   Check the ignition fuse and the fuse connections. If the fuse is blown, replace it with a new one; if the connections are loose or corroded, clean or repair them.
9   Refer to Chapter 8 and check the ignition switch, engine kill switch, neutral switch, clutch switch and sidestand switch.
10   Refer to Section 3 and check the ignition coil primary and secondary resistance.
11   Refer to Section 4 and check the pickup coil resistance.
12   If the preceding checks produce positive results but there is still no spark at the plug, remove the igniter and have it checked by a Yamaha dealer service department or other repair shop equipped with the special tester required.

Engine starts but misfires

13   If the engine starts but misfires, make the following checks before deciding that the ignition system is at fault.
14   The ignition system must be able to produce a spark across a six millimeter (1/4-inch) gap (minimum). A simple test fixture **(see illustration)** can be constructed to make sure the minimum spark gap can be jumped. Make sure the fixture electrodes are positioned six millimeters apart.
15   Connect one of the spark plug wires to the protruding test fixture electrode, then attach the fixture's alligator clip to a good engine ground (earth).
16   Crank the engine over (it will probably start and run on the remaining cylinders) and see if well-defined, blue sparks occur between the test fixture electrodes. If the minimum spark gap test is positive, the ignition coil for that cylinder (and its companion cylinder) is functioning properly. Repeat the check on one of the spark plug wires that is connected to the other coil. If the spark will not jump the gap during either test, or if it is weak (orange colored), refer to Steps 5 through 11 of this Section and perform the component checks described.

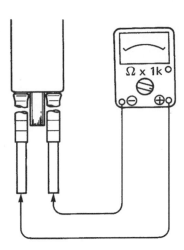

3.6 To test the coil secondary resistance, connect the ohmmeter between the spark plug wires

3.10 To detach the coils, remove the bolt and nut (arrow) - the bolt head may secure a ground (earth) wire

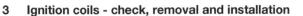

3   Ignition coils - check, removal and installation

Check

Refer to illustrations 3.4 and 3.6

1   In order to determine conclusively that the ignition coils are defective, they should be tested by an authorized Yamaha dealer service department which is equipped with the special electrical tester required for this check.

2   However, the coils can be checked visually (for cracks and other damage) and the primary and secondary coil resistances can be measured with an ohmmeter. If the coils are undamaged, and if the resistances are as specified, they are probably capable of proper operation.

3   To check the coils for physical damage, they must be removed (see Step 9). To check the resistances, remove the seat, fuel tank and body panels as necessary for access (see Chapters 3 and 7). Unplug the primary circuit electrical connectors from the coil(s) and remove the spark plug wires from the plugs that are connected to the coil being checked. Mark the locations of all wires before disconnecting them.

4   To check the coil primary resistance, attach one ohmmeter lead to one of the primary terminals and the other ohmmeter lead to the other primary terminal **(see illustration)**.

5   Place the ohmmeter selector switch in the Rx1 position and compare the measured resistance to the value listed in this Chapter's Specifications.

6   If the coil primary resistance is as specified, check the coil secondary resistance by disconnecting the meter leads from the primary terminals and attaching them to the spark plug wire terminals **(see illustration)**.

7   Place the ohmmeter selector switch in the Rx1000 position and compare the measured resistance to the values listed in this Chapter's Specifications.

8   If the resistances are not as specified, the coil is probably defective and should be replaced with a new one.

Removal and installation

Refer to illustration 3.10

9   If you haven't already done so, remove the seat, the fuel tank and body panels as necessary for access (see Chapters 3 and 7). Look for cylinder number markings on the plug wires and make your own if they aren't visible (number the cylinders one through four, working from the left to the right side of the bike). Disconnect the spark plug wires from the plugs. After labeling them with tape to aid in reinstallation, unplug the coil primary circuit electrical connectors.

4.2 Disconnect the four-terminal connector from the igniter (YX600 Radian shown)

10   Support the coil with one hand and remove the coil mounting bolt and nut **(see illustration)**, then withdraw the coil from its bracket.

11   Installation is the reverse of removal. Make sure the primary circuit electrical connectors and plug wires are attached to the proper terminals.

4   Pickup coils - check, removal and installation

Check

Refer to illustration 4.2

1   Remove the seat (see Chapter 7).

2   If you're working on an early model (with dual pickup coils), disconnect the pickup coil terminal from the igniter **(see illustration)**.

3   If you're working on a later model (with a single pickup coil), follow the wiring harness from the pickup coil(s) to the connector in the wiring harness, then unplug the connector.

4   Probe the terminals in the pickup coil connector with an ohmmeter and compare the resistance reading with the value listed in this Chapter's Specifications. On dual-pickup models, there are two separate tests, between the black terminal and the gray terminal and between the black terminal and the orange terminal.

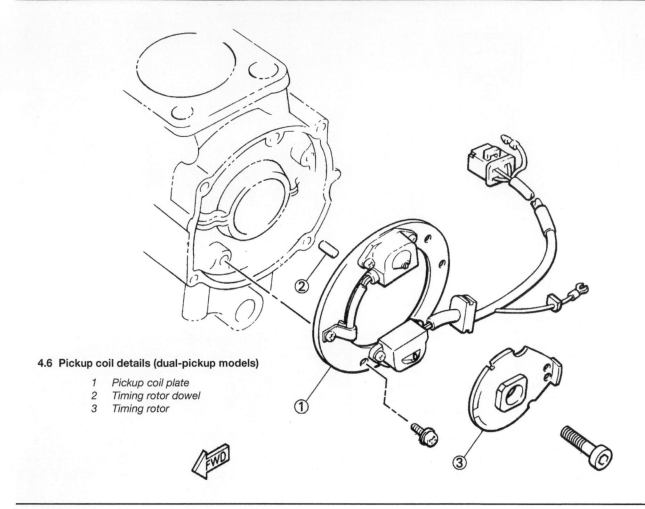

4.6 Pickup coil details (dual-pickup models)

1   *Pickup coil plate*
2   *Timing rotor dowel*
3   *Timing rotor*

5   Set the ohmmeter on the highest resistance range. Measure the resistance between a good ground (earth) and each terminal in the electrical connector. The meter should read infinity. If the pickup coil(s) fail either of the above tests, they must be replaced.

Removal

Dual coil models

Refer to illustration 4.6

6   Remove the timing rotor bolt and take off the rotor **(see illustration)**.
7   Remove the screws that secure the pickup coil plate to the engine case and detach the plate from the engine.
8   Unscrew the pickup coil assembly mounting screws and remove the pickup coils.

Single coil models

9   Remove the alternator cover (see Chapter 8). Remove the stator coil and pickup coil assembly from the cover. The pickup coil isn't available separately and must be replaced together with the stator.

Installation

10   Installation is the reverse of the removal steps, with the following additions:
   a) *If you're working on a dual-pickup model, be sure to seat the grommet on the wiring harness into the notch in the crankcase and route the wiring harness behind the clips on the engine.*

   b) *If you're working on a dual-pickup model, tighten the rotor bolt and plate securing screws to the torques listed in this Chapter's Specifications. Tighten the cover screws to the torque listed in the Chapter 1 Specifications. Use non-permanent thread locking agent on the threads of the plate securing screws.*
   c) *If you're working on a single-pickup model, tighten the stator and pickup coil assembly screws and the alternator cover bolts to the torque listed in the Chapter 8 Specifications.*

---

5   Igniter - check, removal and installation

Check

1   The igniter is checked by process of elimination (when all other possible causes have been checked and eliminated, the igniter is at fault). Because the igniter is expensive and can't be returned once purchased, consider having a Yamaha dealer test the ignition system before you buy a new igniter.

Removal and installation

2   Remove the seat.
3   Unplug the electrical connectors. Remove the mounting screws and take the igniter out **(see illustration 4.2)**.
4   Installation is the reverse of the removal steps.

Chapter 5
Frame, suspension and final drive

Contents

Specifications

Front forks

Fork spring length
FJ600
    Standard .... 472.0 mm (18.6 inches)
    Minimum .... 467.0 mm (18.4 inches)
FZ600
    Standard .... 517.5 mm (20.4 inches)
    Minimum .... 512.5 mm (19.0 inches)
XJ600
    Standard .... 515.5 mm (20.3 inches)
    Minimum .... 510.5 mm (20.1 inches)
YX600 Radian
    Standard .... 542 mm (21.3 inches)
    Minimum .... 537 mm (21.1 inches)
Fork oil capacity, grade and level .... See Chapter 1
Front fork air pressure
Minimum .... 0 Bars (0 psi)
Standard .... 0.39 Bars (5.7 psi)
Maximum .... 0.96 Bars (14 psi)
Maximum difference between forks .... 0.096 Bars (1.4 psi)

Rear suspension

Rear spring free length
    FJ/XJ600 .... 184 mm (7.24 inches)
    FZ600 .... 186 mm (7.32 inches)
    YX600 Radian .... 222.4 mm (8.76 inch)
Swingarm end play and side play limits .... 1 mm (0.040 inch)
Swingarm side clearance (early FJ/XJ600 and
    all YX600 Radian models) .... 0.2 to 0.4 mm (0.008 to 0.0216 inches)
Swingarm bushing length (YX600 Radian models)
    Available thrust washer thicknesses .... 1.9 to 2.0 mm (0.075 to 0.079 inch)
    2.0 to 2.1 mm (0.079 to 0.083 inch)
    2.1 to 2.2 mm (0.083 to 0.087 inch)

Torque specifications

Front forks
    Cap bolt to fork .... 23 Nm (17 ft-lbs)
    Damper rod bolt*
        FJ600 .... Not specified
        FZ600 .... 30 Nm (22 ft-lbs)
        XJ600 .... 23 Nm (17 ft-lbs)
        YX600 Radian .... 30 Nm (22 ft-lbs)
    Drain screw .... 2 Nm (1.4 ft-lbs)
Handlebars and steering stem
    Lower triple clamp bolts.. 23 Nm (17 ft-lbs)
    Upper triple clamp bolts... 20 Nm (14 ft-lbs)
    Handlebar bolts
        FJ/XJ600 .... 70 Nm (51 ft-lbs)
        FZ600 (to fork)... 20 Nm (14 ft-lbs)
        FZ600 (to triple clamp)...................................... 10 Nm (7.2 ft-lbs)
        YX600 Radian .... 20 Nm(14 ft-lbs)
    Steering stem bolt (all except FZ600) .... 54 Nm (39 ft-lbs)
    Steering stem nut (FZ600) .... 110 Nm (80 ft-lbs)
    Steering head bearing ring nut .... See Chapter 1
Rear shock absorber
    Monocross models (FJ600, FZ600, XJ600)
        Upper pivot bolt nut .... 40 Nm (29 ft-lbs)
        Lower pivot bolt nut
            FJ/XJ600 .... 65 Nm (47 ft-lbs)
            FZ600 .... 70 Nm (50 ft-lbs)
    Twin shock models (YX600 Radian)
        Upper shock mount .... 20 Nm (14 ft-lbs)
        Lower shock mount .... 30 Nm (22 ft-lbs)
Rear suspension linkage (Monocross models)
    Tie rods to relay arm .... 65 Nm (47 ft-lbs)
    Tie rod assembly bolts .... 20 Nm (14 ft-lbs)
Swingarm pivot shaft nut... 90 Nm (65 ft-lbs)
Drive chain front sprocket .... 10 Nm (7.2 ft-lbs)
Drive chain rear sprocket .... 32 Nm (23 ft-lbs)*
Engine sprocket cover bolts .... 10 Nm (7.2 ft-lbs)

Use non-permanent thread locking agent on the threads.

1   General information

All motorcycles covered in this manual use a double cradle frame. The frame on FJ600, XJ600 and YX600 Radian models is a one-piece unit constructed of round-section tubing. The FZ600 frame has detachable downtubes and tail section. The FZ600 main section and downtubes are constructed of square-section tubing; the tail section is constructed of round-section tubing.

The front forks are of the conventional coil spring, hydraulically-damped telescopic type. FZ600 forks include air valves so fork air pressure can be adjusted.

The rear suspension on FJ600, FZ600 and XJ600 models is Yamaha's Monocross design, which consists of a single shock absorber with concentric coil spring, a linkage which provides progressive damping and spring rate, and an aluminum swingarm. The rear suspension on YX600 Radian models consists of a swingarm supported by a pair of coil spring/shock absorber units. Rear spring preload is adjustable on all models.

The final drive uses an endless chain (which means it doesn't have a master link). A rubber damper (often called a "cush drive") is installed between the rear wheel coupling and the wheel.

2   Frame - inspection and repair

1   The frame shouldn't require attention unless accident damage has occurred. In most cases, frame replacement is the only satisfactory remedy for such damage. A few frame specialists have the jigs and other equipment necessary for straightening the frame to the required standard of accuracy, but even then there is no simple way of assessing to what extent the frame may have been over-stressed.

2   After the machine has accumulated a lot of miles, the frame should be examined closely for signs of cracking or splitting at the welded joints. Corrosion can also cause weakness at these joints. Loose engine mount bolts can cause ovaling or fracturing of the mounting tabs. Minor damage can often be repaired by welding, depending on the extent and nature of the damage. Welding should be performed by someone expert in the field.

3   Remember that a frame which is out of alignment will cause handling problems. If misalignment is suspected as the result of an accident, it will be necessary to strip the machine completely so the frame can be thoroughly checked.

3   Side and centerstand - maintenance

Refer to illustrations 3.1 and 3.3

1   The centerstand pivots on two bolts attached to the frame **(see illustration)**. Periodically, remove the pivot bolts and grease them thoroughly to avoid excessive wear.

2   Make sure the return spring is in good condition. A broken or weak spring is an obvious safety hazard.

3   The sidestand is attached to a bracket on the frame **(see illustration)**. A pair of extension springs anchored to the bracket ensures that the stand is held in the retracted position.

4   Make sure the pivot bolt is tight and the extension springs are in good condition and not over-stretched. An accident is almost certain to occur if the stand extends while the machine is in motion.

3.1 The centerstand pivots on two bolts

3.3 The sidestand is attached to a bracket on the frame

A   Springs   C   Sidestand switch
B   Pivot   screws

4   Handlebars - removal and installation

Refer to illustrations 4.5, 4.7a and 4.7b

1   The handlebars on the FJ600, FZ600 and XJ600 models are individual units.
2   The handlebar on the YX600 Radian model is a one-piece unit held in place by light alloy caps.
3   If the handlebars must be removed for access to other components, such as the forks or the steering head, it's not necessary to disconnect the cables, wires or brake hose, but it's a good idea to support the assembly with a piece of wire or rope, to avoid unnecessary strain on the cables, wires and the brake hose.
4   If the handlebars are to be removed completely, refer to Chapter 5 for the master cylinder removal procedures, Chapter 3 for the throttle grip removal procedure and Chapter 8 for the switch removal procedure.
5   If you're working on an FJ600 or XJ600, remove the mounting bolt and washer from each handlebar and detach the handlebar **(see illustration).**

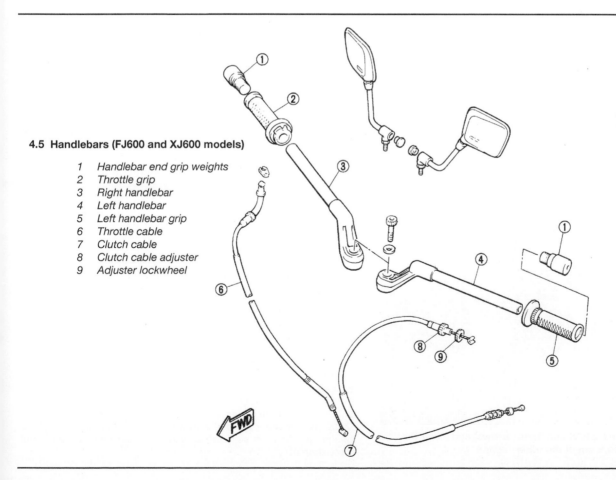

4.5 Handlebars (FJ600 and XJ600 models)

1   Handlebar end grip weights
2   Throttle grip
3   Right handlebar
4   Left handlebar
5   Left handlebar grip
6   Throttle cable
7   Clutch cable
8   Clutch cable adjuster
9   Adjuster lockwheel

4.7a Pry the trim caps from the Allen bolts . . .

4.7b . . . then remove the Allen bolts and lift off the caps to remove the handlebar

6   If you're working on an FZ600, remove two mounting bolts, one horizontal and one vertical, from each handlebar, then lift the handlebar off.

7   If you're working on a YX600 Radian, pry the trim caps out of the handlebar clamps, then unbolt the clamps and lift out the handlebar **(see illustrations).**

8   Check the handlebars for cracks and distortion and replace them if any undesirable conditions are found. When installing the handlebars, tighten the bolts to the torque listed in this Chapter's Specifications and reinstall the trim caps (if equipped).

5   Front forks - removal and installation

Refer to illustration 5.10

Removal

1   Set the bike on its centerstand (if equipped) or support it securely upright.

2   Unbolt the brake hose retainers from the fork legs. Remove the caliper mounting bolts and support the calipers out of the way. Don't disconnect the brake hoses and don't let the calipers hang by the brake hoses.

3   Remove the front wheel (see Chapter 6).

4   Unbolt the front fender and fork brace and take them off.

5.10 Loosen the triple clamp bolts (arrows) and lower the fork out of the triple clamps

FJ/XJ600 models

5   Pry the rubber caps from the fork cap bolts and loosen the cap bolts with an Allen bolt bit.

FZ600 models

6   Remove the master cylinder (see Chapter 6).

7   Remove the handlebars (see Section 4).

8   Hold open the fork air valves until all air pressure is released.

YX600 Radian models

9   Loosen the fork cap bolts with a socket.

All models

10   Loosen the upper and lower triple clamp bolts **(see illustration).** Slide the fork down out of the triple clamps and remove it from the bike.

Installation

11   Installation is the reverse of the removal steps, with the following additions:

a) *Position the ends of the fork tubes flush with the upper triple clamp (FJ600, XJ600 and YX600 Radian) or with the handlebars (FZ600).*

b) *Tighten all fasteners to the torques listed in this Chapter's Specifications.*

c) *If you're working on an FZ600, adjust fork air pressure (see Section 16).*

d) *Pump the front brake lever several times to bring the pads into contact with the discs.*

6   Forks - disassembly, inspection and reassembly

Disassembly

Refer to illustrations 6.2a, 6.2b, 6.2c, 6.4a through 6.4e, 6.5, 6.6, 6.7, 6.8 and 6.9

1   Remove the forks following the procedure in Section 5. Work on one fork leg at a time to avoid mixing up the parts.

2   Remove the fork cap, spring seat and spring **(see illustrations).** (the cap should have been loosened before the forks were removed).

3   Invert the fork assembly over a graduated measuring container and allow the oil to drain out. Note how much fluid was removed. FZ600 models are equipped with a variable damper located under the spring which will slide out as you invert the fork, so be careful not to drop it.

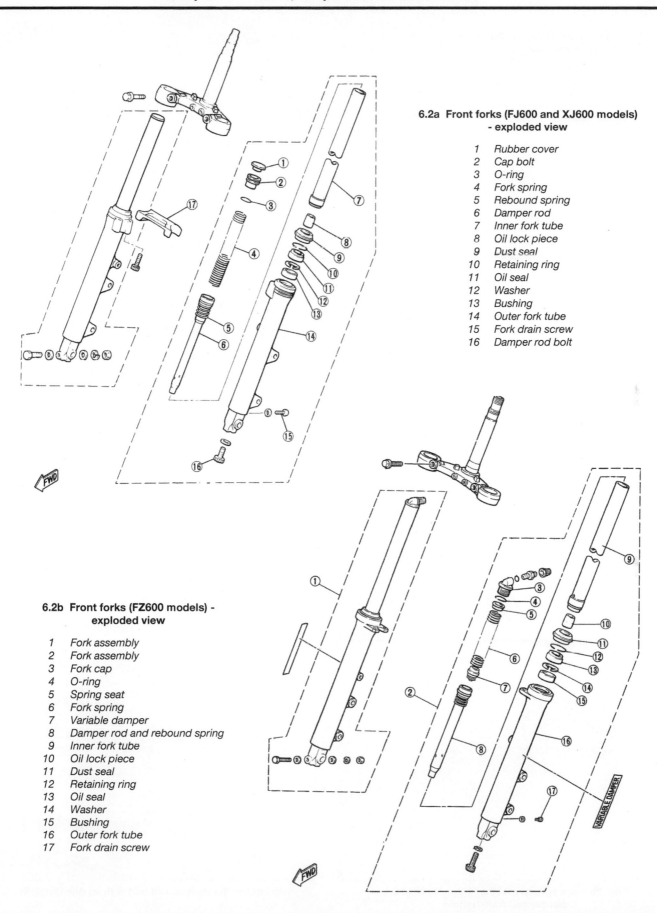

**6.2a Front forks (FJ600 and XJ600 models)
- exploded view**

1	Rubber cover
2	Cap bolt
3	O-ring
4	Fork spring
5	Rebound spring
6	Damper rod
7	Inner fork tube
8	Oil lock piece
9	Dust seal
10	Retaining ring
11	Oil seal
12	Washer
13	Bushing
14	Outer fork tube
15	Fork drain screw
16	Damper rod bolt

**6.2b Front forks (FZ600 models) -
exploded view**

1	Fork assembly
2	Fork assembly
3	Fork cap
4	O-ring
5	Spring seat
6	Fork spring
7	Variable damper
8	Damper rod and rebound spring
9	Inner fork tube
10	Oil lock piece
11	Dust seal
12	Retaining ring
13	Oil seal
14	Washer
15	Bushing
16	Outer fork tube
17	Fork drain screw

6.2c Front forks (YX600 Radian models) -
exploded view

1   Fork assembly
2   Fork assembly
3   Fork cap bolt
4   O-ring
5   Spring seat
6   Fork spring
7   Damper rod and rebound spring
8   Inner fork tube
9   Oil lock piece
10   Dust seal
11   Retaining ring
12   Oil seal
13   Washer
14   Bushing
15   Outer fork tube
16   Fork drain screw

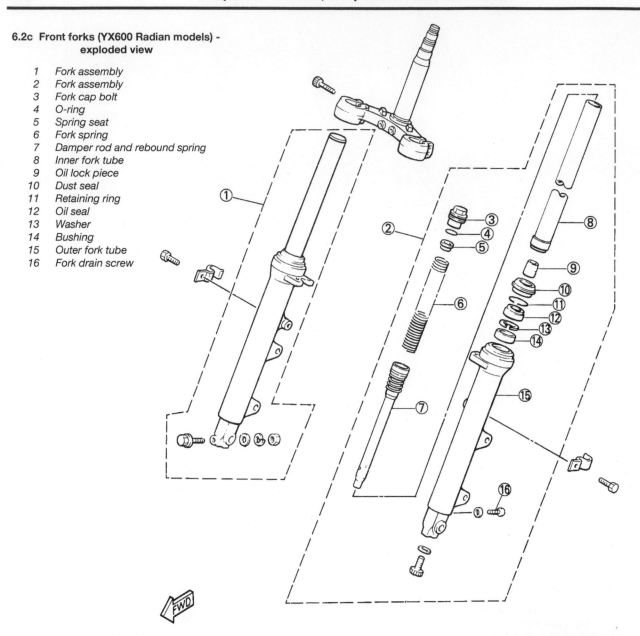

6.4a This is a special tool that's used to hold the
damper rod from turning

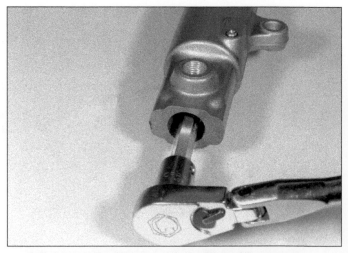

6.4b Loosen the damper rod bolt with an Allen wrench . . .

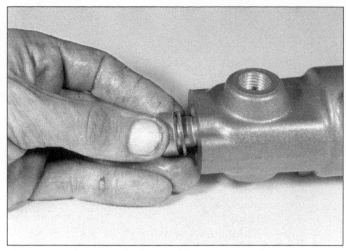

6.4c . . . and remove the bolt and its copper washer - use a new copper washer during reassembly

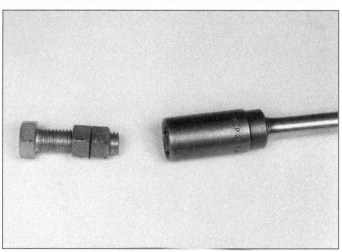

6.4d To make a damper rod holder, thread two nuts onto a bolt with a head that fits inside the damper rod and tighten the nuts against each other . . .

6.4e . . . then install the nut-end into a socket (connected to a long extension) and tape it into place

4   Prevent the damper rod from turning using a holding handle and adapter (Yamaha tools YM-33298 and YM-01326 in the US or 90890-01326 and 90890-04104 in the UK) **(see illustration)**. On later XJ600 models, use the same holding handle with adapter YM-04300-1. Unscrew the Allen bolt at the bottom of the outer tube and retrieve the copper washer **(see illustrations)**. **Note**: *If you don't have access to these special tools, you can fabricate your own using a bolt with a head that fits inside the top of the damper rod in the fork,, two nuts, a socket (to fit on the nuts), a long extension and a ratchet. Thread the two nuts onto the bolt and tighten them against each other* **(see illustration)**. *Insert the assembly into the socket and tape it into place* **(see illustration)**. *Now, insert the tool into the fork tube and engage the bolt head (or the special Yamaha tool) into the hex recess in the damper rod and remove socket head bolt. Another option is to use a piece of hardwood dowel with a taper cut in the end. Push this firmly into the end of the damper rod to keep it from turning.*

5   Tip out the damper rod and the rebound spring **(see illustration)**. Don't remove the Teflon ring from the damper rod unless a new one will be installed.

6   Pry the dust seal from the outer tube **(see illustration)**.

6.5 Remove the damper rod and the Teflon ring - don't separate the ring from the damper rod unless you plan to replace it

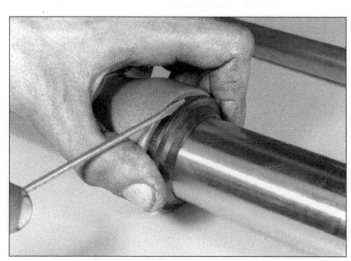

6.6 Pry the dust seal out of the outer tube

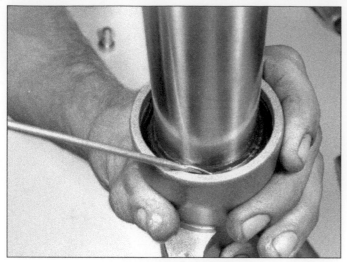

6.7 Pry out the retaining ring

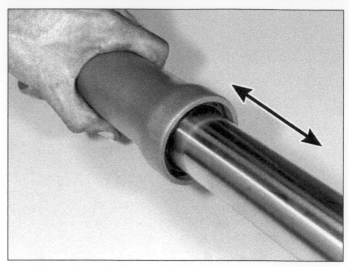

6.8 To separate the inner and outer fork tubes, pull them apart firmly several times - the slide hammer effect will pull the tubes apart

6.9 These parts will come out with the inner fork tube

1	Seal	3	Outer tube guide bushing
2	Washer	4	Inner tube guide bushing

7   Pry the retaining ring from its groove in the outer tube **(see illustration)**. Remove the ring.

8   Hold the outer tube and yank the inner tube away from it, repeatedly (like a slide hammer), until the seal and outer tube guide bushing pop loose **(see illustration)**.

9   Slide the seal, washer and bushing from the inner tube **(see illustration)**.

Inspection

Refer to illustrations 6.11 and 6.12

10   Clean all parts in solvent and blow them dry with compressed air, if available. Check the inner and outer fork tubes, the bushings and the damper rod for score marks, scratches, flaking of the chrome and excessive or abnormal wear. Look for dents in the tubes and replace them if any are found. Check the fork seal seat for nicks, gouges and scratches. If damage is evident, leaks will occur around the seal-to-outer tube junction. Replace worn or defective parts with new ones.

11   On FZ600 models, measure the variable damper height **(see illustration)**. If it's not as listed in this Chapter's Specifications, loosen the locknut and turn the adjuster nut to change it, then retighten the locknut.

12   Check the inner tube for runout with a dial indicator and V-blocks **(see illustration)**. **Warning**: *If a tube is bent, it should not be straightened; replace it with a new one.*

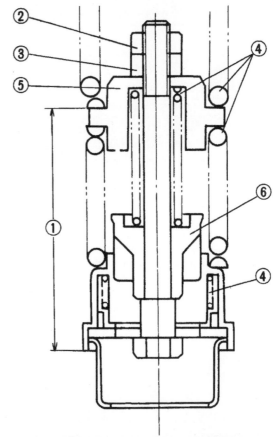

6.11 Adjust the height of the variable damper

1	Damper height	4	Springs
2	Locknut	5	Spring seat
3	Adjuster nut	6	Spool

13   Measure the overall length of the fork spring and check it for cracks and other damage. Compare the length to the minimum length listed in this Chapter's Specifications. If it's defective or sagged, replace both fork springs with new ones. Never replace only one spring!

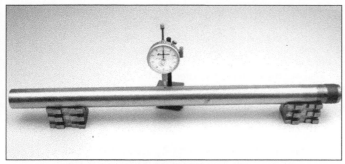

6.12 Check the inner fork tube for runout with a dial indicator and V-blocks

6.17b . . . if you don't have the proper tool, a section of pipe can be used the same way the special tool would be used - as a slide hammer (be sure to tape the ends of the pipe so it doesn't scratch the fork tube)

6.17a Drive the bushing into position with a tool like this one if it's available (use the tool like a slide hammer) . . .

18   Slide the washer down the inner tube, into position over the guide bushing.

19   Lubricate the lips and the outer diameter of the fork seal with the recommended fork oil (see Chapter 1) and slide it down the inner tube, with the lip facing down. Drive the seal into place with the same tools used to drive in the guide bushing. If you don't have access to these, it is recommended that you take the assembly to a Yamaha dealer service department or other motorcycle repair shop to have the seal driven in. If you are very careful, the seal can be driven in with a hammer and a drift punch. Work around the circumference of the seal, tapping gently on the outer edge of the seal until it's seated. Be careful - if you distort the seal, you'll have to disassemble the fork again and end up taking it to a dealer anyway!

20   Install the retaining ring, making sure the ring is completely seated in its groove.

21   Install the dust seal, making sure it seats completely.

22   Install the drain screw and a new gasket, if it was removed.

23   If you're working on an FZ600, install the variable damper with its locknut upward.

24   Add the recommended type and amount of fork oil (see Chapter 1).

25   Install the fork spring, with the closer-wound coils at the top.

26   Install the spring seat on top of the spring.

27   Refer to Chapter 1 and install the fork cap.

28   Install the fork by following the procedure outlined in Section 5. If you won't be installing the fork right away, store it in an upright position.

Reassembly

Refer to illustrations 6.17a and 6.17b

14   If it's necessary to replace the inner guide bushing (the one that won't come off that's on the bottom of the inner tube), pry it apart at the slit and slide it off. Make sure the new one seats properly.

15   Place the rebound spring over the damper rod, then slide the rod assembly into the inner fork tube until it protrudes from the lower end of the tube. Fit the oil lock piece over the end of the damper rod.

16   Insert the inner tube/damper rod assembly into the outer tube until the Allen bolt (with a new copper washer) can be threaded into the damper rod from the lower end of the outer tube. **Note:** *Apply a non-permanent thread locking compound to the threads of the bolt. Using the tool described in Step 4, hold the damper rod and tighten the Allen bolt to the torque listed in this Chapter's Specifications.* **Note:** *If you didn't use the tool, tighten the damper rod bolt after the fork spring and cap bolt are installed.*

17   Slide the outer guide bushing down the inner tube. Using a special bushing driver (Yamaha tool nos. YM-33963 and YM-08010 in the US or 90890-01367 and 90890-01370 in the UK) and a used guide bushing placed on top of the guide bushing being installed, drive the bushing into place until it's fully seated **(see illustration)**. If you don't have access to one of these tools, it is highly recommended that you take the assembly to a Yamaha dealer service department or other motorcycle repair shop to have this done. It is possible, however, to drive the bushing into place using a section of pipe and an old guide bushing **(see illustration)**. Wrap tape around the ends of the pipe to prevent it from scratching the fork tube.

7   Steering head bearings - replacement

Removal

1   If the steering head bearing check/adjustment (see Chapter 1) does not remedy excessive play or roughness in the steering head bearings, the entire front end must be disassembled and the bearings and races replaced with new ones.

2   Remove the handlebars (see Section 4), the front wheel (see Chapter 6) and the front forks (see Section 5).

FJ600 and XJ600 models

3   Remove the upper fairing (see Chapter 7).

4   Remove the instrument cluster (see Chapter 8).

FZ600 models

5   Remove the upper, center and lower fairings (see Chapter 7).

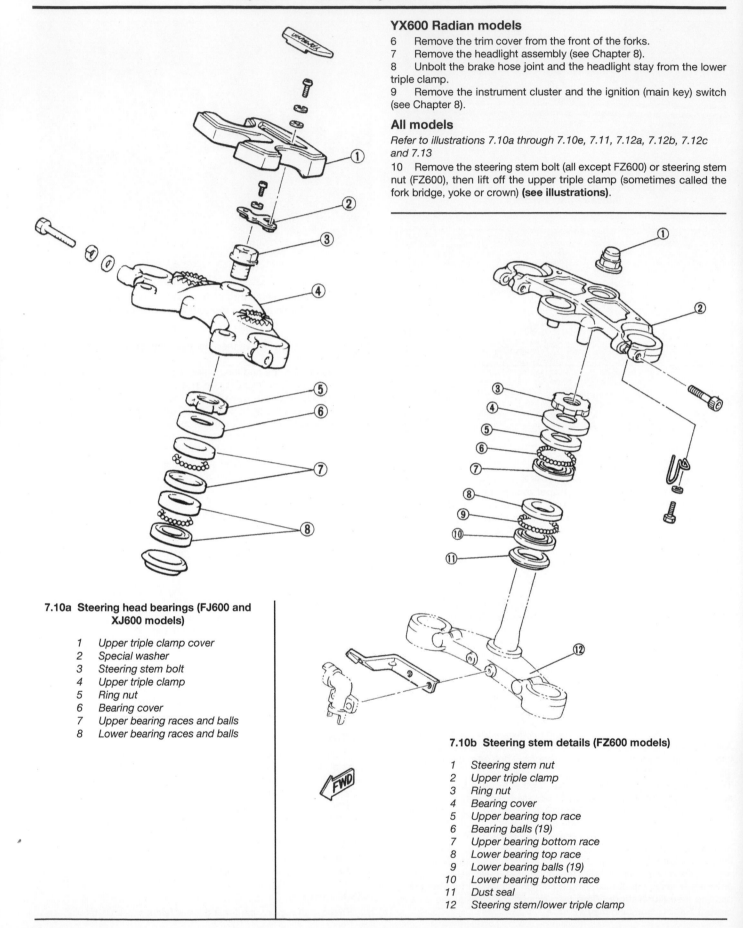

YX600 Radian models

6   Remove the trim cover from the front of the forks.
7   Remove the headlight assembly (see Chapter 8).
8   Unbolt the brake hose joint and the headlight stay from the lower triple clamp.
9   Remove the instrument cluster and the ignition (main key) switch (see Chapter 8).

All models

Refer to illustrations 7.10a through 7.10e, 7.11, 7.12a, 7.12b, 7.12c and 7.13

10   Remove the steering stem bolt (all except FZ600) or steering stem nut (FZ600), then lift off the upper triple clamp (sometimes called the fork bridge, yoke or crown) **(see illustrations)**.

7.10a Steering head bearings (FJ600 and XJ600 models)

1   Upper triple clamp cover
2   Special washer
3   Steering stem bolt
4   Upper triple clamp
5   Ring nut
6   Bearing cover
7   Upper bearing races and balls
8   Lower bearing races and balls

7.10b Steering stem details (FZ600 models)

1   Steering stem nut
2   Upper triple clamp
3   Ring nut
4   Bearing cover
5   Upper bearing top race
6   Bearing balls (19)
7   Upper bearing bottom race
8   Lower bearing top race
9   Lower bearing balls (19)
10   Lower bearing bottom race
11   Dust seal
12   Steering stem/lower triple clamp

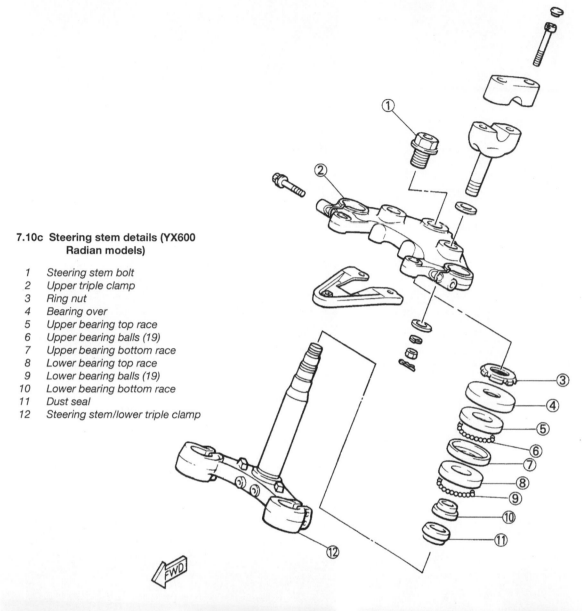

7.10c Steering stem details (YX600 Radian models)

1   Steering stem bolt
2   Upper triple clamp
3   Ring nut
4   Bearing over
5   Upper bearing top race
6   Upper bearing balls (19)
7   Upper bearing bottom race
8   Lower bearing top race
9   Lower bearing balls (19)
10 Lower bearing bottom race
11 Dust seal
12 Steering stem/lower triple clamp

7.10d Unscrew the steering stem bolt or nut . . .

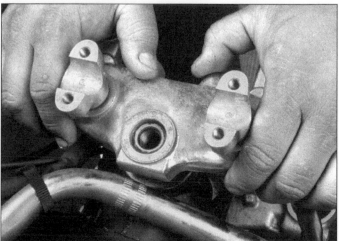

7.10e . . . and remove the upper triple clamp (YX600 Radian shown)

7.11 Unscrew the ring nut

7.12a Lift off the bearing cover . . .

7.12b . . . and the upper bearing top race, taking care not to let the bearing balls drop out . . .

7.12c . . . and remove the bearing balls

7.13 Carefully lower the steering stem out of the steering head and remove the bearing balls

7.16 Drive the lower bearing top race out from above; drive the upper bearing bottom race out from below

11   Remove the ring nut from the steering stem **(see illustration)**.

12   Remove the bearing cover, upper bearing race and 19 steel bearing balls **(see illustrations)**. Take care not to let the steel balls drop, as they are easily lost.

13   Lower the steering stem out of the steering head, again taking care not to lose any of the bearing balls **(see illustration)**. If the steering stem is stuck, gently tap on the top end with a plastic mallet or a hammer and wood block.

Inspection

Refer to illustrations 7.16 and 7.20

14   Clean all the parts with solvent and dry them thoroughly, using compressed air, if available. Wipe the old grease out of the frame steering head and bearing races.

15   Examine the races in the steering head for cracks, dents, and pits. If even the slightest amount of wear or damage is evident, the races should be replaced with new ones.

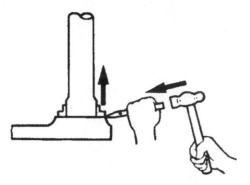

7.20 Tap a chisel between the lower bearing bottom race and the lower triple clamp to remove the race

16   To remove the bearing races, drive them out of the steering head with a hammer and punch **(see illustration)**. A slide hammer with the proper internal-jaw puller will also work.

17   Since the races are installed with an interference fit in the frame, installation will be easier if the new races are left overnight in a freezer. This will cause them to contract and slip into place in the frame with very little effort. When installing the races, tap them gently into place with a hammer and punch or a large socket. Do not strike the bearing surface or the race will be damaged.

7.22 Stick the bearing balls in position with grease

18   Check the bearings for wear. Look for cracks, dents, and pits in the races and flat spots, pitting or galling on the bearing balls. Replace any defective parts with new ones. If a new bearing is required, replace both bearings and their races as a set.

19   Check the dust seal under the lower bearing and replace it with a new one if necessary.

20   Don't remove the lower bearing bottom race unless it must be replaced. To remove the bottom race, tap between the race and steering stem with a chisel **(see illustration)**. The seal will be ruined during this process, so replace it with a new one.

21   Inspect the steering stem/lower triple clamp for cracks and other damage. Do not attempt to repair any steering components. Replace them with new parts if defects are found.

Installation

Refer to illustration 7.22

22   Pack the lower bearing bottom race with high-quality wheel bearing grease. Stick 19 steel balls to the grease around the race, then add more grease **(see illustration)**. **Note:** *A small hand-operated grease gun will make this job easier. Coat the upper and lower races with grease also.*

23   Pack the upper bearing and install the steel balls in the same way as for the lower bearing.

24   Slip the steering stem/lower triple clamp into the steering head, taking care not to dislodge any of the bearing balls.

25   Install the top race on top of the upper bearing balls, then install the bearing cover and ring nut. Install the forks, then refer to Chapter 1 and adjust the bearings.

26   The remainder of installation is the reverse of the removal steps.

---

8   Rear shock absorber - removal, inspection and installation

---

Removal

Warning: *The shock absorber on FJ/XJ600 and FZ600 models is filled with highly pressurized nitrogen gas. Be extremely careful when discarding the shock unit. Refer to safe disposal instructions at the end of this Section.*

1   Set the bike on its centerstand (if equipped) or support it securely upright.

FJ600 and XJ600 models

Refer to illustrations 8.5a and 8.5b

2   Remove the exhaust system (see Chapter 3).

3   Remove the seat and side covers (see Chapter 7).

4   Remove the battery (see Chapter 8), then remove the battery box.

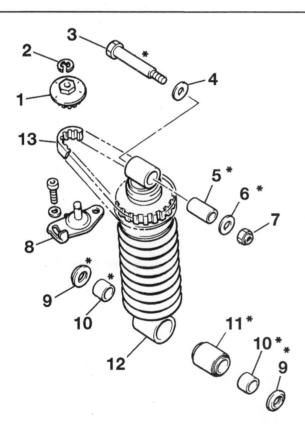

8.5a Rear shock absorber details (FJ600 and early XJ600 models)

1	*Damping adjuster knob*	7	*Locknut*
2	*C-clip*	8	*Damping adjuster*
3	*Shock absorber upper*	9	*Dust cover*
	bolt	10	*Collar*
4	*Washer*	11	*Bushing*
5	*Bushing*	12	*Shock absorber*
6	*Washer*	13	*Damping adjuster belt*

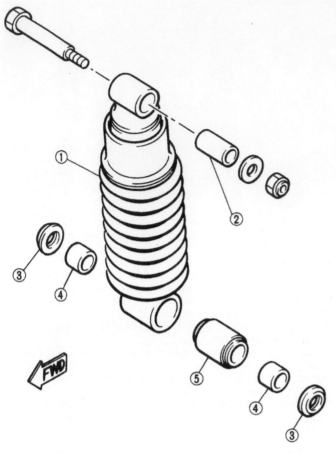

8.5b Rear shock absorber details (later XJ600 models)

1	Shock absorber	4	Collar
2	Bushing	5	Bushing
3	Dust cover		

5   Remove the shock absorber lower bolt and its dust seals and collars **(see illustrations)**.

6   Remove the upper mounting bolt and nut. On models so equipped, disengage the damper adjusting belt.

7   Remove the shock absorber from the motorcycle.

FZ600 models

Refer to illustration 8.11

8   Remove the lower and center fairings, seat and side covers (see Chapter 7).

9   Remove the exhaust system (see Chapter 3).

10   Remove the lower fairing stays. Support the engine with a jack.

11   Remove the shock absorber lower bolt and its dust seals and collars **(see illustration)**.

12   Remove the remote control unit and adjusting belt.

13   Remove the shock absorber upper nut and bolt and remove the shock absorber from the motorcycle.

YX600 Radian models

Refer to illustrations 8.15a, 8.15b and 8.15c

14   Remove the mufflers (see Chapter 3).

15   Remove the shock absorber mounting bolts and take the shock absorber out **(see illustrations)**.

Inspection

Refer to illustration 8.19

16   Check the shock for signs of oil or gas leaks and replace it if you find any.

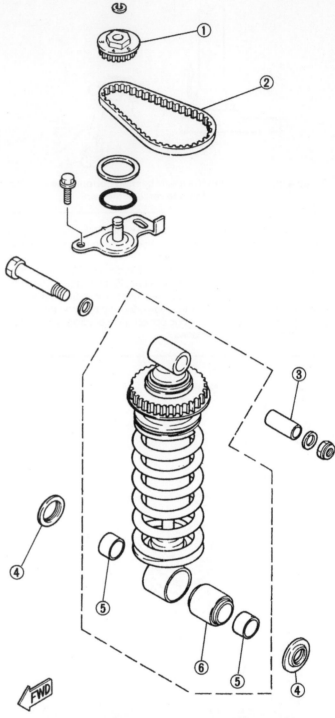

8.11 Rear shock absorber details (FZ600 models)

1	Damping adjuster knob	4	Dust cover
2	Damping adjuster belt	5	Collar
3	Bushing	6	Bushing

17   Inspect the pivot hardware at the top and bottom of the shock and replace any worn or damaged parts.

18   Check the remote control unit and adjusting belt (if equipped) for wear, damage or rough movement. Replace them if any defects are found.

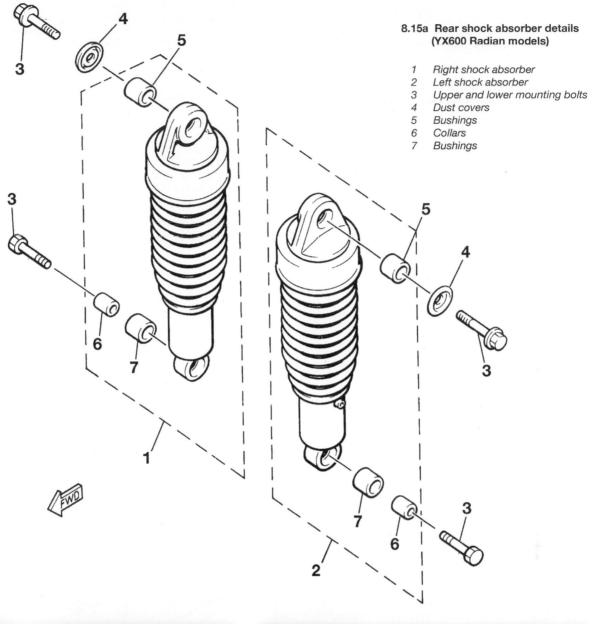

8.15a Rear shock absorber details
(YX600 Radian models)

1   Right shock absorber
2   Left shock absorber
3   Upper and lower mounting bolts
4   Dust covers
5   Bushings
6   Collars
7   Bushings

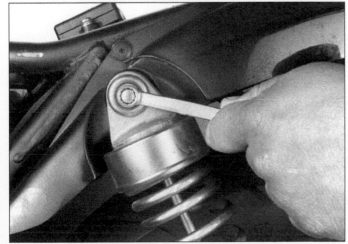

8.15b Remove the upper mounting bolt . . .

8.15c . . . and the lower mounting bolt and lift the
shock absorber out

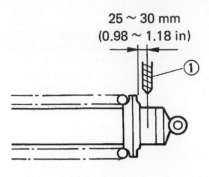

25 ~ 30 mm
(0.98 ~ 1.18 in)

8.19 Wear eye protection and drill a hole at the point shown to release gas from the shock absorber before discarding it

19   Before you discard a worn shock absorber on FJ600, FZ600 or XJ600 models, release the gas pressure by drilling a 2-3 mm (0.08-0.12 in) hole through the top of the shock cylinder wall **(see illustration)**. **Warning:** *Wear eye protection to prevent eye damage from escaping gas and/or metal chips*. You can also take the shock absorber to a Yamaha dealer for safe disposal.

Installation

20   Installation is the reverse of the removal steps, with the following additions:

a) *Apply multi-purpose lithium grease to the collars and pivot bolt shafts.*

b) *Tighten all fasteners to the torques listed in this Chapter's Specifications except the remote preload adjuster mounting bolts (if equipped).*

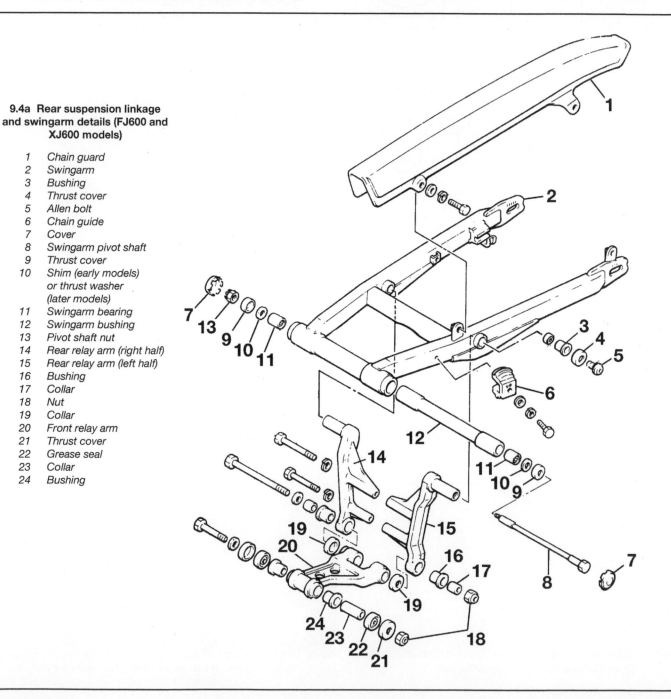

9.4a Rear suspension linkage and swingarm details (FJ600 and XJ600 models)

1   Chain guard
2   Swingarm
3   Bushing
4   Thrust cover
5   Allen bolt
6   Chain guide
7   Cover
8   Swingarm pivot shaft
9   Thrust cover
10 Shim (early models)
    or thrust washer
    (later models)
11 Swingarm bearing
12 Swingarm bushing
13 Pivot shaft nut
14 Rear relay arm (right half)
15 Rear relay arm (left half)
16 Bushing
17 Collar
18 Nut
19 Collar
20 Front relay arm
21 Thrust cover
22 Grease seal
23 Collar
24 Bushing

c) *If you're working on an FJ600, FZ600 or early XJ600 with a remote adjuster for the rear spring preload setting, tighten the remote adjuster bolts slightly, so the remote adjuster is free to pivot around the bolts. Attach a spring scale and pull it outward with a force of 20 kg (44 lbs) to adjust the belt tension. Tighten the remote adjuster bolts to the torque listed in this Chapter's Specifications, then release the spring scale.*

9 Rear suspension linkage - removal, inspection and installation

Refer to illustrations 9.4a and 9.4b

1   Set the bike on its centerstand (if equipped) or prop it securely upright.

2   Disconnect the lower end of the rear shock absorber (see Section 8).

3   Remove exhaust system components and body panels as needed for access (see Chapters 3 and 7).

4   Remove the pivot bolt from the relay arm and take the arm out **(see illustrations).**

5   Lower the rear relay arm and remove the two bolt(s) holding the halves of the relay arm together. Remove the screws that secure the halves of the relay arm to the swingarm, then separate the halves and remove them from the swingarm.

6   Remove the thrust covers and bushing sleeves. Inspect the covers, sleeves, pivot bolts and bushings and replace them if they're worn or damaged.

7   Installation is the reverse of the removal steps, with the following additions:

a) *Lubricate the bushings, pivot bolts and thrust covers with lithium-base waterproof wheel bearing grease.*

b) *Tighten the nuts and bolts to the torques listed in this chapter's Specifications.*

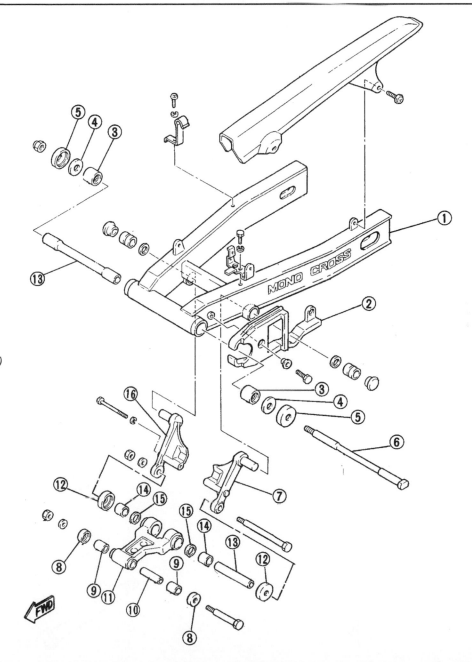

9.4b Rear suspension linkage and swingarm details (FZ600 models)

1   Swingarm
2   Chain guide
3   Swingarm bearing
4   Plate washer
5   Thrust cover
6   Swingarm pivot shaft
7   Rear relay arm (left half)
8   Thrust cover
9   Collar
10   Bushing
11   Front relay arm
12   Thrust cover
13   Collar
14   Bushing
15   Grease seal
16   Rear relay arm (right half)

11.8 Remove the nut (arrow) from the swingarm pivot shaft

10 Swingarm bearings - check

1   Refer to Chapter 6 and remove the rear wheel, then refer to Section 8 and remove the rear shock absorber.

2   Grasp the rear of the swingarm with one hand and place your other hand at the junction of the swingarm and the frame. Try to move the rear of the swingarm from side-to-side. Any wear (play) in the bearings should be felt as movement between the swingarm and the frame at the front. The swingarm will actually be felt to move forward and backward at the front (not from side-to-side). If any play is noted, the bearings should be replaced with new ones (see Section 12).

3   Next, move the swingarm up and down through its full travel. It should move freely, without any binding or rough spots. If it doesn't move freely, refer to Section 11 for servicing procedures.

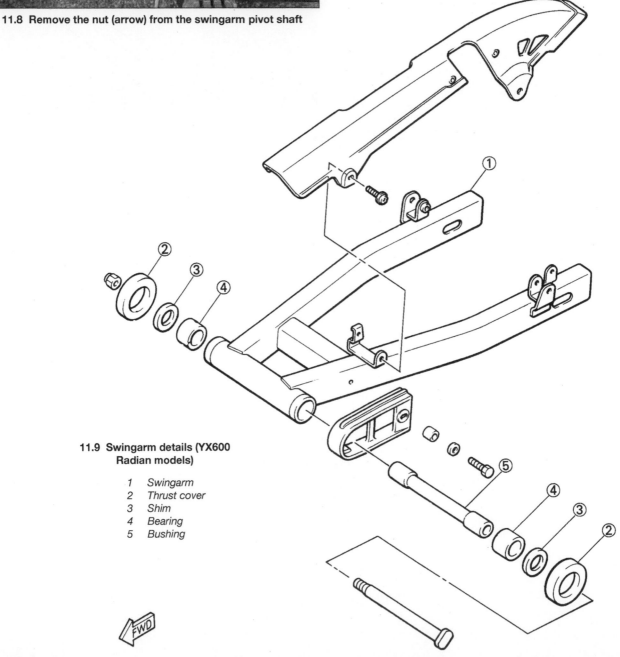

11.9 Swingarm details (YX600 Radian models)

1   Swingarm
2   Thrust cover
3   Shim
4   Bearing
5   Bushing

12.2 Pry off the thrust cover and remove the washer . . .

12.3 . . . then pull out the bushing . . .

12.4 . . . to expose the bearings

11 Swingarm - removal and installation

Refer to illustrations 11.8 and 11.9

1 Set the bike on its centerstand (if equipped) or prop it securely upright.
2 If you're working on a YX600 Radian, remove the mufflers (see Chapter 3).
3 Remove the rear wheel (see Chapter 6).
4 Remove any support brackets and detach associated wiring, cables, or hoses from the swingarm.
5 Detach the rear brake caliper on models so equipped (see Chapter 6).
6 Detach the brake torque link from the swingarm.
7 Detach the shock absorber(s) and rear suspension linkage (if equipped) from the swingarm (see Section 9).
8 Remove the swingarm pivot nut and washer **(see illustration)**.
9 Support the swingarm and pull the pivot shaft out **(see illustration 9.4a, 9.4b or the accompanying illustration)**. Remove the swingarm. If necessary, remove the bolts and detach the suspension linkage relay arm from the swingarm.
10 Check the pivot bearings in the swingarm for dryness or deterio-

ration. If they're in need of lubrication or replacement, refer to Section 12.
11 Installation is the reverse of the removal procedure. Be sure the bearing seals are in position before installing the pivot shaft. Tighten all fasteners to the torques listed in this Chapter's Specifications. Adjust the chain as described in Chapter 1.

12 Swingarm bearings - inspection and replacement

Refer to illustrations 12.2, 12.3, 12.4, 12.5 and 12.6

1 Remove the swingarm (see Section 11).
2 Remove the dust cap and thrust washer from each side of the swingarm **(see illustration)**.
3 Slide the bushing out **(see illustration)**.
4 Inspect the bearings **(see illustration)**. If they're dry, lubricate them with lithium base waterproof wheel bearing grease. If they're worn or damaged, take the swingarm to a Yamaha dealer or motorcycle repair shop for bearing replacement.
5 If you're working on a YX600 Radian, measure the length of the swingarm bushing **(see illustration)**. If it's not within the range listed in this Chapter's Specifications, replace the bushing.

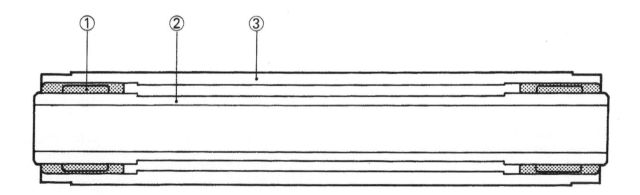

12.5 Measure the length of the swingarm bushing

1 Bearing *2 Bushing* *3 Swingarm*

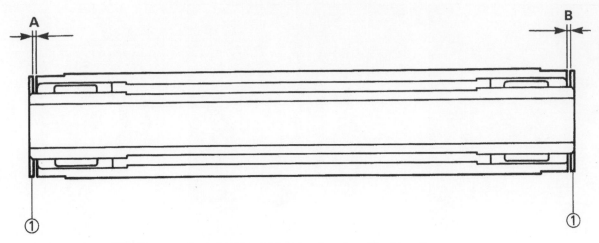

12.6 Measure the swingarm side play and replace the shims if necessary

A   Side clearance   B   Side clearance   1   Shims

**13.2 Remove the Allen bolts (arrows) and take off the
engine sprocket cover**

6   If you're working on an FJ600, an early XJ600 or a YX600 Radian, calculate the swingarm side clearance. Install the bushing in the swingarm so it's perfectly centered **(see illustration)**. Install the shims so they're flush with the ends of the bushing and measure the gaps between the shims and swingarm. The combined total of the two gaps is swingarm side clearance. If it's not the same as the value listed in this Chapter's Specifications, install thicker or thinner shims to correct it. Available shim thicknesses are listed in this Chapter's Specifications.

---

13   Drive chain - removal, cleaning and installation

Refer to illustration 13.2

Removal

Warning: *The original equipment chain has no master link. Don't cut the chain and install a master link or it may break while the motorcycle is in motion, possibly causing rear wheel lockup and loss of control of the motorcycle.*

1   Disconnect the shift pedal arm from its shaft on the engine (see Chapter 2).
2   Remove the bolts securing the engine sprocket cover to the engine case. Take the sprocket cover off **(see illustration)**.

3   Remove the rear wheel (see Chapter 6).
4   Lift the chain off the engine sprocket.
5   Detach the swingarm from the frame by following the first few Steps of Section 11. Pull the swingarm back far enough to allow the chain to slip between the frame and the front of the swingarm.

Cleaning

6   Soak the chain in kerosene (paraffin) for approximately five or six minutes. **Caution:** *Don't use gasoline (petrol), solvent or other cleaning fluids. Don't use high-pressure water. Remove the chain, wipe it off, then blow dry it with compressed air immediately. The entire process shouldn't take longer than ten minutes - if it does, the O-rings in the chain rollers could be damaged.*

Installation

7   Installation is the reverse of the removal procedure. Tighten the suspension fasteners and the engine sprocket cover bolts to the torque values listed in this Chapter's Specifications. Tighten the rear axle nut to the torque listed in the Chapter 6 Specifications.
8   Lubricate the chain following the procedure described in Chapter 1. **Caution:** *Use only the recommended engine oil.*

---

14   Sprockets - check and replacement

1   Set the bike on its centerstand (if equipped) or support it securely upright with the rear wheel off the ground.
2   Whenever the drive chain is inspected, the sprockets should be inspected also. If you are replacing the chain, replace the sprockets as well. Likewise, if the sprockets are in need of replacement, install a new chain also.
3   Remove the engine sprocket cover (see Section 13).
4   Check the wear pattern on the sprockets (see Chapter 1). If the sprocket teeth are worn excessively, replace the chain and sprockets.
5   To remove the rear sprocket, remove the rear wheel (see Chapter 6). Unscrew the Allen bolts and nuts holding the sprocket to the wheel coupling and lift it off. Check the condition of the rubber damper under the rear wheel coupling (see Section 15).
6   To replace the engine sprocket, shift the transmission into gear. Remove the two bolts and the mounting plate.
7   Pull the engine sprocket and chain off the shaft, then separate the sprocket from the chain.
8   When installing the engine sprocket, engage it with the chain. Install the mounting plate and tighten the bolts to the torque listed in this Chapter's Specifications.

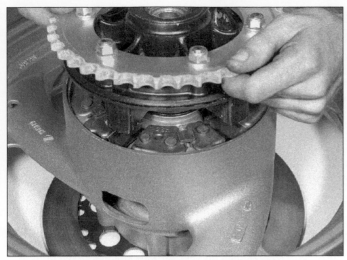

15.2 Lift the sprocket out of the hub

15.3 Lift out the rubber damper segments

	Front fork Air pressure	Rear shock absorber Spring seat	Loading condition			
			Solo rider	With passenger	With accessories and equipment	With accessories, equipment and passenger
1.	39.2 ~ 58.9 kPa (0.4 ~ 0.6 kg/cm , 5.7 ~ 8.5 psi)	1 ~ 2	O			
2.	39.2 ~ 59.0 kPa (0.4 ~ 0.6 kg/cm , 5.7 ~ 8.5 psi)	3 ~ 5		O		
3.	58.9 ~ 78.5 kPa (0.6 ~ 0.8 kg/cm , 8.5 ~ 11 psi)	3 ~ 5			O	
4.	78.5 kPa (0.8 kg/cm², 11 psi)	5				O

16.3 Suspension settings (FZ600 models)

9   Install the engine sprocket cover and the shift pedal (see Chapter 2).
10   Use non-permanent thread locking agent on the threads of the rear sprocket bolts.

15 Rear wheel coupling/rubber damper - check and replacement

Refer to illustrations 15.2 and 15.3
1   Remove the rear wheel (see Chapter 6).
2   Lift the collar and rear sprocket/rear wheel coupling from the wheel **(see illustration)**.
3   Lift the rubber damper segments from the wheel **(see illustration)** and check them for cracks, hardening and general deterioration. Replace the rubber dampers as a set if necessary.
4   Checking and replacement procedures for the coupling bearing are similar to those described for the wheel bearings. Refer to Chapter 6.
5   Installation is the reverse of the removal procedure.

16 Suspension adjustments

1   Front fork air pressure on FZ600 models can be adjusted for different riding conditions.
2   The rear spring preload on all models can be adjusted for different riding conditions.

FZ600 models

Refer to illustrations 16.3 and 16.6
3   Select a combination of fork air pressure and rear spring preload from the accompanying chart **(see illustration).**
4   Remove the cap from each front fork.
5   Place an air pressure gauge onto the air valve and measure pressure. Increase pressure to stiffen the forks and decrease pressure to soften them. **Caution:** *Maximum air pressure is only 0.96 Bars (14 psi), and exceeding the maximum may damage the fork oil seals. Use a hand pump to add air. Don't use a service station air hose.*
Warning: *Make sure air pressure in both forks is even within the limit listed in this Chapter's Specifications or unstable handling may occur.*

16.6 Align the number on the damper knob with the indictor

1   *Indicator*

6   Adjust rear spring preload by turning the adjuster on the right side of the motorcycle. Remove the front and rear seats and the right side cover (see Chapter 7). Turn the adjuster with the special wrench included in the bike's tool kit **(see illustration).**

FJ/XJ600 and YX600 Radian models

Refer to illustration 16.10

7   The front forks are not adjustable.

8   Rear spring preload on 1984 through 1987 models is adjusted with a knob on the right side of the motorcycle similar to that used on FZ600 models **(see illustration 16.6).**

9   Rear spring preload on 1989 and later XJ600 models is adjusted by turning the ring at the top of the shock absorber tool kit. Turn the

16.10 On YX600 Radian models, turn the adjuster at the bottom of the shock to set preload

adjuster clockwise to increase spring preload and counterclockwise (anti-clockwise) to decrease it. The softest setting is 1; standard is 3; the hardest setting is 7.

10   Rear spring preload on YX600 Radian models is adjusted by turning the adjuster at the bottom of the shock absorber **(see illustration). Warning:** *Be sure to set both shock absorbers to the same setting or unstable handling may occur.*

Chapter 6
Brakes, wheels and tires

Contents

Specifications

Brakes

Brake fluid type	See Chapter 1
Brake pad thickness	See Chapter 1
Brake disc thickness (front and rear)	
Standard	5.0 mm (0.20 inch)
Minimum*	4.0 mm (0.14 inch)
Disc runout limit	not specified
Brake drum diameter	
Standard	180 mm (7.08 inch)
Maximum	181 mm (7.12 inch)
Brake shoe lining thickness	
Standard	4 mm (0.16 inch)
Minimum*	2 mm (0.08 inch)

*Refer to marks stamped into the disc (they supersede information printed here)

Wheels and tires

Wheel runout limit (all except US FZ600)	2.0 mm (0.08 inch)
Wheel runout limit (US FZ600)	
Vertical	1.0 mm (0.04 inch)
Lateral	0.5 mm (0.02 inch)
Tire pressures and tread depth	See Chapter 1
Tire sizes**	
FJ600	
Front	90/90-18 51H
Rear	120/80-18 62H
FZ600	
Front	100/90-16 54H
Rear	120/80-18 62H
XJ600	
Front	90/90-18 51H
Rear	110/90-18 61H
YX600 (Radian)	
Front	110/90-16 59H
Rear	130/90-16 67H

**Refer to the tire information/fitment label on the motorcycle (it supersedes information printed here).

Torque specifications

Caliper mounting bolts	35 Nm (25 ft-lbs)
Front axle	Not available
Front axle clamp bolt and nut	20 Nm (14 ft-lbs)
Brake disc mounting bolts..	20 Nm (14 ft-lbs)
Union (banjo fitting) bolts	26 Nm (19 ft-lbs)
Bleeder screws	6 Nm (4.3 ft-lbs)
Master cylinder mounting bolts	
Front	
FJ600	8.5 NM (6.1 ft-lbs)
XJ600	9 Nm (6.5 ft-lbs)
FZ600 and YX600 Radian	8 Nm (4.3 ft-lbs)
Rear	20 Nm (14 ft-lbs)
Rear axle nut..	105 Nm (75 ft-lbs)
Torque link bolts	
FZ600	26 Nm (19 ft-lbs)
YX600 Radian...	20 Nm (14 ft-lbs)
Caliper bracket bolts (FJ600 and XJ600)	35 Nm (25 ft-lbs)

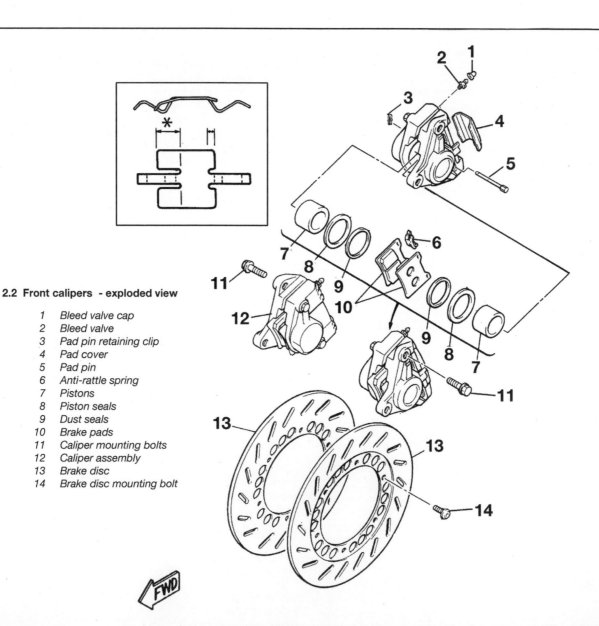

2.2 Front calipers - exploded view

1   *Bleed valve cap*
2   *Bleed valve*
3   *Pad pin retaining clip*
4   *Pad cover*
5   *Pad pin*
6   *Anti-rattle spring*
7   *Pistons*
8   *Piston seals*
9   *Dust seals*
10   *Brake pads*
11   *Caliper mounting bolts*
12   *Caliper assembly*
13   *Brake disc*
14   *Brake disc mounting bolt*

2.9a Special lubricants are required in the UK to prevent corrosion

A   *Apply Duckhams Copper 10 or equivalent to the shaded areas*
B   *Apply Shin-Etsu G-40M or equivalent silicone grease to the shaded areas*

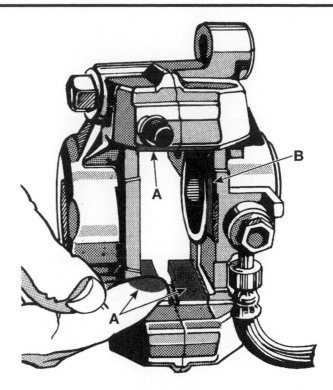

2.9b Apply the recommended lubricants to the pad friction areas inside the caliper and to the exposed portion of the caliper pistons

A   *Duckhams Copper 10*
B   *Shin-Etsu G-40M or equivalent silicone grease*

1   General information

All models covered by this manual use a pair of fixed-mount dual piston calipers at the front. FJ600, FZ600 and XJ600 models use one fixed-mount dual piston caliper at the rear. YX600 Radian models use mechanically actuated drum brakes at the rear.

All models are equipped with cast aluminum wheels, which require very little maintenance and allow tubeless tires to be used.

Caution: *Disc brake components rarely require disassembly. Do not disassemble components unless absolutely necessary. If any hydraulic brake line connection in the system is loosened, the entire system should be disassembled, drained, cleaned and then properly filled and bled upon reassembly. Do not use solvents on internal brake components. Solvents will cause seals to swell and distort. Use only clean brake fluid, brake cleaner or alcohol for cleaning. Use care when working with brake fluid as it can injure your eyes and it will damage painted surfaces and plastic parts.*

2   Brake pads - replacement

Front calipers

Refer to illustrations 2.2, 2.9a and 2.9b
Warning: *When replacing the front brake pads always replace the pads in BOTH calipers - never just on one side. Also, the dust created by the brake system may contain asbestos, which is harmful to your health. Never blow it out with compressed air and don't inhale any of it. An approved filtering mask should be worn when working on the brakes.*

1   Set the bike on its centerstand if equipped) or prop it securely upright.
2   Remove the brake pad cover from each caliper **(see illustration)**.
3   Pull the retaining clips out of the pad pins.

4   Pull out one of the pad pins.
5   Remove the pad spring plate. Note that the longer end of the spring faces forward (in the rotating direction of the brake discs).
6   Pull out the remaining pad pin and lift the pads out of the caliper.
7   Check the condition of the brake discs (see Section 4). If they're in need of machining or replacement, follow the procedure in that Section to remove them. If they're okay, deglaze them with sandpaper or emery cloth, using a swirling motion.
8   Remove the cover and diaphragm from the master cylinder reservoir and siphon out some fluid. Push the pistons into the caliper as far as possible, while checking the master cylinder reservoir to make sure it doesn't overflow. If you can't depress the pistons with thumb pressure, try using a C-clamp (G-clamp). If the piston sticks, remove the caliper and overhaul it as described in Section 3.
9   **Warning**: *This step is necessary to ensure that the pads move freely in the calipers. Because a large amount of salt is used on roads in the UK, special lubrication of the pads and calipers is required. Where this is the case, apply a thin film of Duckhams Copper 10 or equivalent to the following areas before installing the pads* **(see illustrations)**:
   a) *To the edges of the metal backing on the brake pads*
   b) *To the pad retaining pins*
   c) *To the areas of the caliper where the pads rub*
   d) *To the threads of the caliper mounting bolts.*
Apply a thin film of Shin-Etsu G-40M or equivalent silicone grease to the following:
   a) *Exposed areas of the caliper pistons*
   b) *The areas of the pad backing plates that contact the pistons.*
Caution: *Don't use too much Copper 10 and don't apply it to the pad pin retaining clips or the anti-rattle shim(s). Make sure no Copper 10 contacts the brake discs or the pad friction surfaces.*
10   Install the pads in the caliper and position the pad spring on the pads with its longer end facing forward.

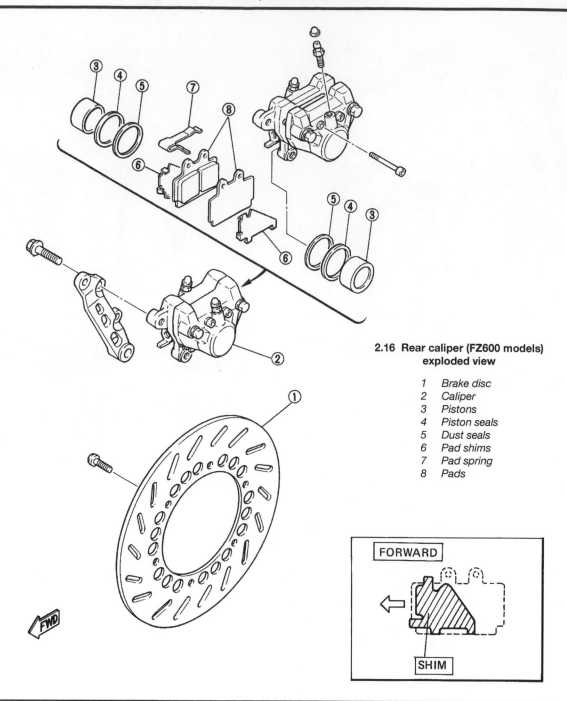

**2.16 Rear caliper (FZ600 models)
exploded view**

1   Brake disc
2   Caliper
3   Pistons
4   Piston seals
5   Dust seals
6   Pad shims
7   Pad spring
8   Pads

FORWARD

SHIM

11   Slide the pad pins into their holes, making sure they pass over the fingers on the pad spring.
12   Install the retaining clips in the pad pins.
13   Refill the master cylinder reservoir (see Chapter 1) and install the diaphragm and cap.
14   Operate the brake lever several times to bring the pads into contact with the disc. Check the operation of the brakes carefully before riding the motorcycle.

Rear calipers

Warning: *The dust created by the brake system may contain asbestos, which is harmful to your health. Never blow it out with compressed air and don't inhale any of it. An approved filtering mask should be worn when working on the brakes.*

15   Set the bike on its centerstand (if equipped) or support it securely upright.

FZ600 models

Refer to illustration 2.16

16   Loosen the pad pins, but don't remove them yet (they screw into the caliper) **(see illustration)**.
17   Remove the caliper mounting bolt and lift the caliper off the disc.
18   Unscrew the pad pins and pull them out. Remove the pads, shims and pad spring from the bottom of the caliper.
19   Perform Steps 7, 8 and 9 above to inspect the disc and prepare the caliper for pad installation.
20   Install the spring, pads and shims in the caliper **(see illustration 2.16)**.Be sure the spring and shims face in the proper direction.
21   Install the pad pins and tighten them finger-tight.
22   Install the caliper and tighten its mounting bolts to the torque listed in this chapter's Specifications. Tighten the pad pins securely.

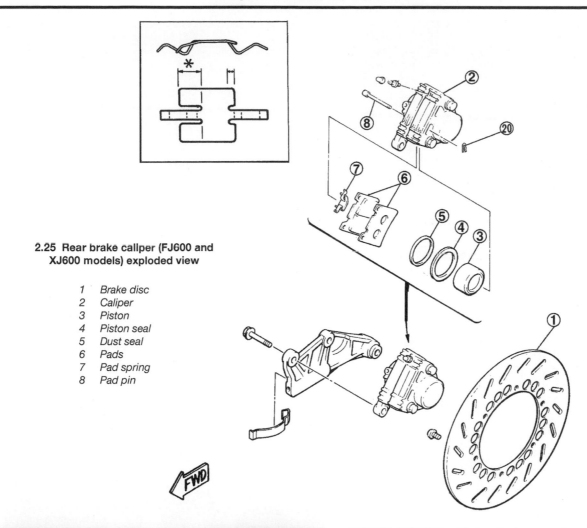

2.25 Rear brake caliper (FJ600 and XJ600 models) exploded view

1   Brake disc
2   Caliper
3   Piston
4   Piston seal
5   Dust seal
6   Pads
7   Pad spring
8   Pad pin

3.2a Remove the union bolt and two sealing washers - this is a front caliper . . .

A   Union bolt   B   Caliper mounting bolts

FJ600 and XJ600 models

Refer to illustration 2.25

23   If you're working on a model with a pad cover on top of the caliper, remove it.
24   If you're working on a model without a pad cover, remove the caliper mounting bolts and lift the caliper and pads off the disc.
25   Remove the clips from the pad pins and pull the pad pins out. Remove the pads and pad spring **(see illustration).**

26   Perform Steps 7, 8 and 9 above to inspect the disc and prepare the caliper for pad installation.
27   Installation is the reverse of the removal steps. Be sure the long end of the pad spring faces forward **(see illustration 2.25)**. If you removed the caliper, tighten its mounting bolts to the torque listed in this chapter's Specifications.

3   Brake caliper - removal, overhaul and installation

Warning: *If a front caliper indicates the need for an overhaul (usually due to leaking fluid or sticky operation), BOTH front calipers should be overhauled and all old brake fluid flushed from the system. Also, the dust created by the brake system may contain asbestos, which is harmful to your health. Never blow it out with compressed air and don't inhale any of it. An approved filtering mask should be worn when working on the brakes. Do not, under any circumstances, use petroleum-based solvents to clean brake parts. Use brake cleaner or denatured alcohol only!*

Removal

Refer to illustrations 3.2a and 3.2b

1   Place the bike on its centerstand (if equipped) or prop it securely upright.
2   Disconnect the brake hose from the caliper. Remove the brake hose banjo fitting bolt and separate the hose from the caliper **(see illustrations)**. Discard the sealing washers. Place the end of the hose in a container and operate the brake lever to pump out the fluid. Once this is done, wrap a clean shop rag tightly around the hose fitting to soak up any drips and prevent contamination.

3.2b ... and this is a rear caliper; be sure to use new sealing washers on installation

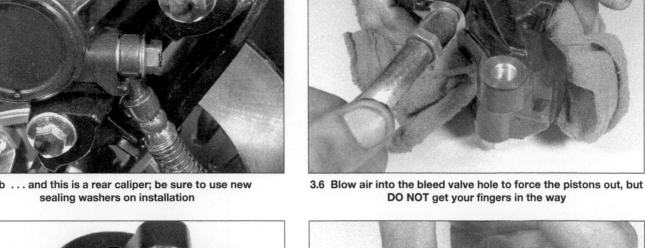

3.6 Blow air into the bleed valve hole to force the pistons out, but **DO NOT** get your fingers in the way

3.8a Each caliper bore contains a dust seal and a piston seal

1 Dust seal	2 Piston seal

3.8b Remove the dust seal from its groove . . .

3   Unscrew the caliper mounting bolts **(see illustrations 2.2, 2.16 and 2.25)**. **Caution**: *Don't remove the bolts that hold the caliper halves together.*

4   Lift off the caliper.

Overhaul

Refer to illustrations 3.6, 3.8a, 3.8b and 3.9

5   Clean the exterior of the caliper with denatured alcohol or brake system cleaner.

6   Place a few rags between the pistons and the caliper frame to act as a cushion, then use compressed air, directed into the fluid inlet, to remove the pistons **(see illustration)**. Use only enough air pressure to ease the pistons out of the bore. If a piston is blown out, even with the cushion in place, it may be damaged. **Warning**: *Never place your fingers in front of the piston in an attempt to catch or protect it when applying compressed air, as serious injury could occur.* If you Don't have a compressor, a service station air hose will work. If the pistons won't all come out, push the pistons that do come out back into their bores and hold them there with C-clamps (G-clamps) while you blow out the remaining piston.

7   If compressed air isn't available, reconnect the caliper to the brake hose and pump the brake lever or pedal until the pistons are free.

8   Using a wood or plastic tool, remove the dust seals **(see illustrations)**. Metal tools may cause bore damage.

9   Using a wood or plastic tool, remove the piston seals from the groove in the caliper bore **(see illustration)**.

10   Clean the pistons and the bores with denatured alcohol, clean brake fluid or brake system cleaner and blow dry them with filtered,

3.9 ... then remove the piston seal

4.3 Set up a dial indicator to contact each brake disc, then rotate the wheel to check for runout

4.6 Loosen the mounting bolts evenly to detach the brake disc from the wheel

unlubricated compressed air. Inspect the surfaces of the pistons for nicks and burrs and loss of plating. Check the caliper bores, too. If surface defects are present, the caliper must be replaced. If the caliper is in bad shape, the master cylinder should also be checked.

11   Lubricate the piston seals with clean brake fluid and install it in its groove in the caliper bore. Make sure it isn't twisted and seats completely.

12   Lubricate the dust seals with clean brake fluid and install them in their grooves, making sure they seat correctly.

13   Lubricate the pistons with clean brake fluid and install them into the caliper bores. Using your thumbs, push the pistons all the way in, making sure they don't get cocked in the bore.

Installation

14   Install the caliper, tightening the mounting bolts to the torque listed in this chapter's Specifications.

15   Connect the brake hose to the caliper, using new sealing washers on each side of the fitting. Tighten the banjo fitting bolt to the torque listed in this chapter's Specifications.

16   Fill the master cylinder with the recommended brake fluid (see Chapter 1) and bleed the system (see Section 8). Check for leaks.

17   Check the operation of the brakes carefully before riding the motorcycle.

---

4   Brake disc(s) - inspection, removal and installation

Inspection

Refer to illustration 4.3

1   Set the bike on its centerstand (if equipped) or prop it securely upright.

2   Visually inspect the surface of the disc(s) for score marks and other damage. Light scratches are normal after use and Won't affect brake operation, but deep grooves and heavy score marks will reduce braking efficiency and accelerate pad wear. If the discs are badly grooved they must be machined or replaced.

3   To check disc runout, mount a dial indicator to a fork leg or the swingarm, with the plunger on the indicator touching the surface of the disc about 1/2-inch (13 mm) from the outer edge **(see illustration)**. Support the wheel being checked off the ground and slowly turn the wheel (if you're checking the front discs on a model equipped with a centerstand, have an assistant sit on the seat to raise the front wheel off the ground) and watch the indicator needle., comparing your reading with the limit listed in this chapter's Specifications. If the runout is greater than allowed, check the hub bearings for play (see Chapter 1). If the bearings are

worn, replace them and repeat this check. If the disc runout is still excessive, it will have to be replaced.

4   The disc must not be machined or allowed to wear down to a thickness less than the minimum allowable thickness, listed in this chapter's Specifications (check also for wear limits stamped on the disc itself). The thickness of the disc can be checked with a micrometer. If the thickness of the disc is less than the minimum allowable, it must be replaced.

Removal

Refer to illustration 4.6

5   Remove the wheel (see Section 12 for front wheel removal or Section 13 for rear wheel removal). **Caution:** *Don't lay the wheel down and allow it to rest on one of the discs - the disc could become warped. Set the wheel on wood blocks so the disc doesn't support the weight of the wheel.*

6   Mark the relationship of the disc to the wheel, so it can be installed in the same position. Remove the Allen bolts that retain the disc to the wheel **(see illustration)**. Loosen the bolts a little at a time, in a criss-cross pattern, to avoid distorting the disc.

7   Take note of any paper shims that may be present where the disc mates to the wheel. If there are any, mark their position and be sure to include them when installing the disc.

Installation

8   Position the disc on the wheel, aligning the previously applied matchmarks (if you're reinstalling the original disc). Make sure the arrow (stamped on the disc) marking the direction of rotation is pointing in the proper direction.

9   Apply a non-hardening thread locking compound to the threads of the bolts. Install the bolts, tightening them a little at a time, in a criss-cross pattern, until the torque listed in this chapter's Specifications is reached. Clean off all grease from the brake disc(s) using acetone or brake system cleaner.

10   Install the wheel.

11   Operate the brake lever or pedal several times to bring the pads into contact with the disc. Check the operation of the brakes carefully before riding the motorcycle.

---

5   Front brake master cylinder - removal, overhaul and installation

---

1   If the master cylinder is leaking fluid, or if the lever does not produce a firm feel when the brake is applied, and bleeding the brakes does not help, master cylinder overhaul is recommended. Before disassembling the master cylinder, read through the entire procedure

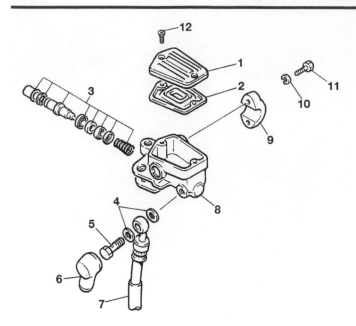

5.3 Front brake master cylinder - exploded view

1	Reservoir cover	6	Rubber cover
2	Diaphragm	7	Brake hose
3	Piston assembly and	8	Master cylinder body
	spring	9	Clamp
4	Brake hose sealing	10	Lockwasher
	washers	11	Allen bolt
5	Union bolt	12	Cover screw

5.6 Remove the locknut and pivot bolt (master cylinder removed for clarity)

5.7 Remove the master cylinder mounting bolts

and make sure that you have the correct rebuild kit. Also, you will need some new, clean brake fluid of the recommended type, some clean rags and internal snap-ring pliers. **Note:** *To prevent damage to the paint from spilled brake fluid, always cover the fuel tank when working on the master cylinder.*

2   **Caution:** *Disassembly, overhaul and reassembly of the brake master cylinder must be done in a spotlessly clean work area to avoid contamination and possible failure of the brake hydraulic system components.*

Removal

Refer to illustrations 5.3, 5.6 and 5.7

3   Loosen, but do not remove, the screws holding the reservoir cover in place **(see illustration)**.

4   Disconnect the electrical connectors from the brake light switch (see Chapter 8).

5   Pull back the rubber boot, loosen the banjo fitting bolt and separate the brake hose from the master cylinder. Wrap the end of the hose in a clean rag and suspend the hose in an upright position or bend it down carefully and place the open end in a clean container. The objective is to prevent excessive loss of brake fluid, fluid spills and system contamination.

6   Remove the locknut from the underside of the lever pivot bolt, then unscrew the bolt **(see illustration)**.

7   Remove the master cylinder mounting bolts **(see illustration)** and separate the master cylinder from the handlebar. **Caution:** *Do not tip the master cylinder upside down or brake fluid will run out.*

Overhaul

Refer to illustrations 5.8, 5.9, 5.10a, 5.10b, 5.10c, 5.11a and 5.11b

8   Detach the reservoir cover and the rubber diaphragm, then drain the brake fluid into a suitable container. Remove the splash plate from the bottom of the reservoir (if equipped) **(see illustration)**, then wipe any remaining fluid out of the reservoir with a clean rag.

9   Carefully remove the rubber dust boot from the end of the piston **(see illustration)**.

10   Using snap-ring pliers, remove the snap-ring **(see illustration)**

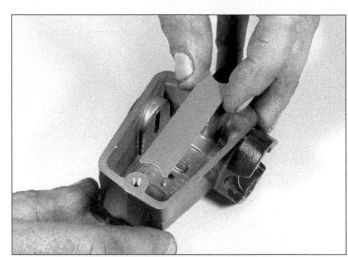

5.8 Lift the splash plate (if equipped) out of the cylinder body so you can clean the fluid ports in the bottom

5.9 Remove the rubber boot

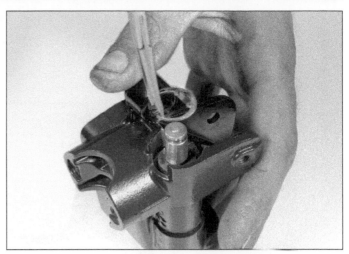

5.10a Remove the snap-ring from the cylinder bore . . .

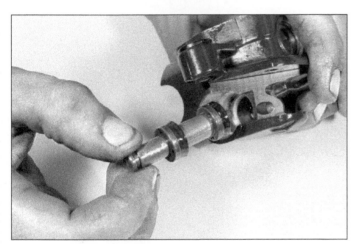

5.10b . . . then pull out the piston assembly . . .

parts. If compressed air is available, use it to dry the parts thoroughly (make sure it's filtered and unlubricated). Check the master cylinder bore for corrosion, scratches, nicks and score marks. If damage is evident, the master cylinder must be replaced with a new one. If the master cylinder is in poor condition, then the calipers should be checked as well.

13   The dust seal, piston assembly and spring are included in the rebuild kit. Use all of the new parts, regardless of the apparent condition of the old ones.

14   Before reassembling the master cylinder, soak the piston and the rubber cup seals in clean brake fluid for ten or fifteen minutes. Lubricate the master cylinder bore with clean brake fluid, then carefully insert the piston and related parts in the reverse order of disassembly. Make sure the lips on the cup seals do not turn inside out when they are slipped into the bore.

15   Depress the piston, then install the snap-ring (make sure the snap-ring is properly seated in the groove). Install the rubber dust boot (make sure the lip is seated properly in the piston groove).

Installation

16   Attach the master cylinder to the handlebar and tighten the bolts to the torque listed in this chapter's Specifications.

Caution: *don't mix up the clutch master cylinder clamp and the brake master cylinder clamp. The clutch master cylinder clamp is slightly bigger and the clamps or cylinders may be damaged if the clamps are installed on the wrong cylinders.*

17   Connect the brake hose to the master cylinder, using new sealing washers. Tighten the banjo fitting bolt to the torque listed in this chapter's Specifications. Refer to Section 8 and bleed the air from the system.

and slide out the piston assembly and the spring **(see illustrations)**. Lay the parts out in the proper order to prevent confusion during reassembly.

11   Remove the spring and pivot bushing from the lever **(see illustrations)**.

12   Clean all of the parts with brake system cleaner (available at auto parts stores), isopropyl alcohol or clean brake fluid. **Caution**: *Do not, under any circumstances, use a petroleum-based solvent to clean brake*

5.10c . . . and spring

5.11a Pull the spring out of the lever . . .

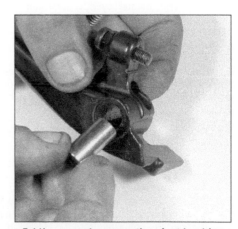

5.11b . . . and remove the pivot bushing

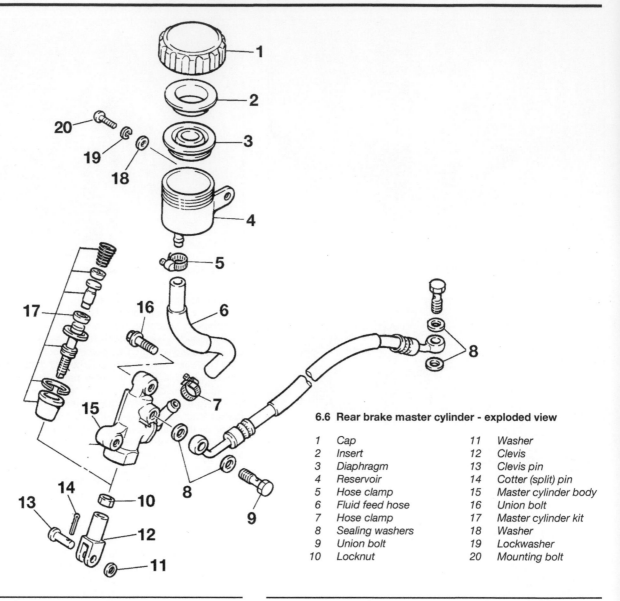

6.6 Rear brake master cylinder - exploded view

1	Cap	11	Washer
2	Insert	12	Clevis
3	Diaphragm	13	Clevis pin
4	Reservoir	14	Cotter (split) pin
5	Hose clamp	15	Master cylinder body
6	Fluid feed hose	16	Union bolt
7	Hose clamp	17	Master cylinder kit
8	Sealing washers	18	Washer
9	Union bolt	19	Lockwasher
10	Locknut	20	Mounting bolt

6   Rear brake master cylinder - removal, overhaul and installation

1   If the master cylinder is leaking fluid, or if the pedal does not produce a firm feel when the brake is applied, and bleeding the brakes does not help, master cylinder overhaul is recommended. Before disassembling the master cylinder, read through the entire procedure and make sure that you have the correct rebuild kit. Also, you will need some new, clean brake fluid of the recommended type, some clean rags and internal snap-ring pliers.

2   **Caution**: *Disassembly, overhaul and reassembly of the brake master cylinder must be done in a spotlessly clean work area to avoid contamination and possible failure of the brake hydraulic system components.*

Removal

Refer to illustration 6.6

3   Set the bike on its centerstand (if equipped) or prop it securely upright.

4   If you're working on an FJ600 or XJ600, remove the right side cover (see Chapter 7) and the right footpeg bracket.

5   If you're working on an FZ600, remove the protective cover.

6   Remove the cotter pin from the clevis pin on the master cylinder pushrod **(see illustration)**. Remove the clevis pin.

7   Have a container and some rags ready to catch spilling brake fluid. Using a pair of pliers, slide the clamp up the fluid feed hose and detach the hose from the master cylinder. Direct the end of the hose into the container, unscrew the cap on the master cylinder reservoir and allow the fluid to drain.

8   Using a six-point box-end wrench (ring spanner), unscrew the banjo union bolt from the master cylinder. Discard the sealing washers on each side of the fitting.

Overhaul

9   Remove the screw, lockwasher and washer and detach the fluid inlet fitting from the master cylinder. Remove the O-ring from the bore.

10   Hold the clevis with a pair of pliers and loosen the locknut. Unscrew the clevis and locknut from the pushrod and carefully remove the rubber dust boot from the pushrod.

11   Depress the pushrod and, using snap-ring pliers, remove the snap-ring. Slide out the piston assembly and spring. Lay the parts out in the proper order to prevent confusion during reassembly.

12   Clean all of the parts with brake system cleaner (available at auto parts stores), isopropyl alcohol or clean brake fluid. **Caution**: *Do not, under any circumstances, use a petroleum-based solvent to clean brake parts. If compressed air is available, use it to dry the parts thoroughly (make sure it's filtered and unlubricated). Check the master cylinder bore for corrosion, scratches, nicks and score marks. If damage is evident, the*

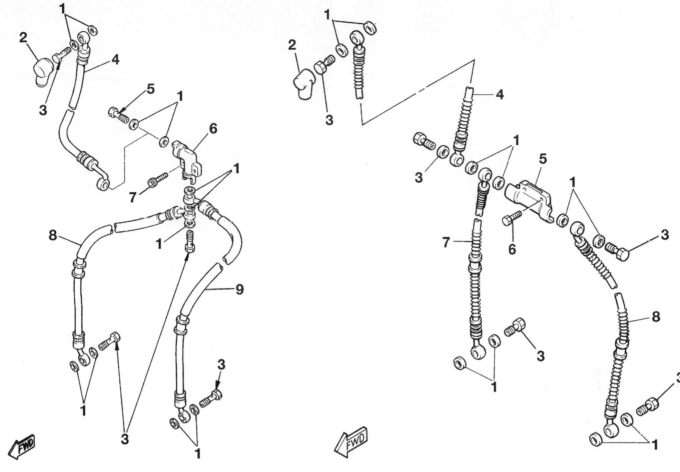

7.2a Front brake hose details (FJ600, FZ600 and XJ600)

1   Sealing washers
2   Rubber cap
3   Union bolt
4   Brake hose from master cylinder
5   Union bolt
6   Brake hose union
7   Union mounting bolt
8   Hose to right caliper
9   Hose to left caliper

7.2b Front brake hose details (YX600 Radian)

1   Sealing washers
2   Rubber cap
3   Union bolts
4   Hose from master cylinder
5   Hose union
6   Union mounting bolt
7   Hose to right caliper
8   Hose to left caliper

master cylinder must be replaced with a new one. If the master cylinder is in poor condition, then the caliper should be checked as well.

13   A new piston and spring are included in the rebuild kit. Use them regardless of the condition of the old ones.

14   Before reassembling the master cylinder, soak the piston and the rubber cup seals in clean brake fluid for ten or fifteen minutes. Lubricate the master cylinder bore with clean brake fluid, then carefully insert the parts in the reverse order of disassembly. Make sure the lips on the cup seals do not turn inside out when they are slipped into the bore.

15   Depress the pushrod, then install the snap-ring (make sure the snap-ring is properly seated in the groove). Install the rubber dust boot (make sure the lip is seated properly in the groove).

16   Install the clevis to the end of the pushrod, then tighten the locknut.

Installation

17   Connect the banjo fitting to the top of the master cylinder, using a new sealing washer on each side of the fitting. Tighten the banjo fitting bolt to the torque listed in this chapter's Specifications.

18   Connect the fluid feed hose to the inlet fitting and install the hose clamp.

19   Connect the clevis to the brake pedal and secure the clevis pin with a new cotter pin.

20   Fill the fluid reservoir with the specified fluid (see Chapter 1) and bleed the system following the procedure in Section 8. Install the right side cover (see Chapter 7).

21   Check the position of the brake pedal (see Chapter 1) and adjust it if necessary. Check the operation of the brakes carefully before riding the motorcycle.

7   Brake hoses - inspection and replacement

Inspection

Refer to illustrations 7.2a and 7.2b

1   Once a week, or if the motorcycle is used less frequently, before every ride, check the condition of the brake hoses.

2   Twist and flex the rubber hoses **(see illustration 6.6 and the accompanying illustrations)** while looking for cracks, bulges and seeping fluid. Check extra carefully around the areas where the hoses connect with the banjo fittings, as these are common areas for hose failure.

8.6a Each front caliper has a single bleed valve (arrow) . . .

8.6b . . . and the rear caliper has two bleed valves (arrows)

3 Inspect the metal banjo union fittings connected to brake hoses. If the plating on the metal tubes is chipped or scratched, the lines may rust. If the fittings are rusted, scratched or cracked, replace them.

Replacement

4 Most brake hoses have banjo union fittings on each end of the hose. Cover the surrounding area with plenty of rags and unscrew the banjo bolt on each end of the hose. Detach the hose from any clips that may be present and remove the hose.
5 Position the new hose, making sure it isn't twisted or otherwise strained, between the two components. Make sure the metal tube portion of the banjo fitting is located between the protrusions on the component it's connected to, if equipped. In some cases, there's a cast lug (on caliper bodies, for example) that keeps the brake hose from spinning clockwise as the banjo bolt is tightened. Install the banjo bolts, using new sealing washers on both sides of the fittings, and tighten them to the torque listed in this chapter's Specifications.
6 Flush the old brake fluid from the system, refill the system with the recommended fluid (see Chapter 1) and bleed the air from the system (see Section 8). Check the operation of the brakes carefully before riding the motorcycle.

8 Brake system bleeding

Refer to illustrations 8.6a, 8.6b, 8.8a and 8.8b

1 Bleeding the brake is simply the process of removing all the air bubbles from the brake fluid reservoirs, the lines and the brake calipers. Bleeding is necessary whenever a brake system hydraulic connection is loosened, when a component or hose is replaced, or when the master cylinder or caliper is overhauled. Leaks in the system may also allow air to enter, but leaking brake fluid will reveal their presence and warn you of the need for repair.
2 Bleed the brake in the following order: right front caliper, left front caliper and rear caliper.
3 To bleed the brakes, you will need some new, clean brake fluid of the recommended type (see Chapter 1), a length of clear vinyl or plastic tubing, a small container partially filled with clean brake fluid, some rags and a wrench to fit the brake caliper bleeder valves.
4 Cover the fuel tank and other painted components to prevent damage in the event that brake fluid is spilled.
5 Remove the reservoir cap or cover and slowly pump the brake lever or pedal a few times, until no air bubbles can be seen floating up from the holes at the bottom of the reservoir. Doing this bleeds the air from the master cylinder end of the line. Reinstall the reservoir cap or cover.

8.8a Loosen the bleed valve to release fluid (this is the front caliper) . . .

6 Slip a box wrench (ring spanner) over the caliper bleed valve **(see illustrations)**. Attach one end of the clear vinyl or plastic tubing to the bleeder valve and submerge the other end in the brake fluid in the container.
7 Remove the reservoir cap or cover and check the fluid level. Do not allow the fluid level to drop below the lower mark during the bleeding process.
8 Carefully pump the brake lever or pedal three or four times and hold it. Open the caliper bleed valve **(see illustrations)**. When the valve is opened, brake fluid will flow out of the caliper into the clear tubing and the lever will move toward the handlebar or the pedal will move down.
9 Retighten the bleeder valve, then release the brake lever or pedal gradually. Repeat the process until no air bubbles are visible in the brake fluid leaving the caliper and the lever or pedal is firm when applied. **Note**: *On rear calipers with two bleed valves, air must be bled from both, one after the other. Remember to add fluid to the reservoir as the level drops. Use only new, clean brake fluid of the recommended type. Never reuse the fluid lost during bleeding.*
10 Replace the reservoir cover or cap, wipe up any spilled brake fluid and check the entire system for leaks. **Note**: *If bleeding is difficult, it may be necessary to let the brake fluid in the system stabilize for a few hours (it may be aerated). Repeat the bleeding procedure when the tiny bubbles in the system have settled out.*

8.8b . . . and this is the rear caliper

9.4 Lift the brake panel out of the drum

9.5 Measure the lining thickness

9.8 Measure drum diameter

9 Rear drum brakes - removal, inspection and installation

Removal

Refer to illustration 9.4

1   Before you start, inspect the brake wear indicator (see Chapter 1).
2   Place the motorcycle onto its centerstand.
3   Remove the rear wheel (see Section 13).
4   Lift the brake panel from the brake drum built into the wheel **(see illustration 13.6c and the accompanying illustration)**.

Inspection

Refer to illustrations 9.5 and 9.8

Warning: *The dust created by the brake system may contain asbestos, which is harmful to your health. Never blow it out with compressed air and don't inhale any of it. An approved filtering mask should be worn when working on the brakes. Do not, under any circumstances, use petroleum-based solvents to clean brake parts. Use brake cleaner or denatured alcohol only!*

5   Measure the thickness of the lining material(just the lining material, not the metal backing) and compare with the value listed in this chapter's Specifications **(see illustration)**. Replace the shoes if the material is worn to less than the minimum. Also check the linings

for wear, damage and signs of contamination from road dirt or water. If the linings are visibly defective, replace them.

6   Fold the shoes toward each other to release the spring tension. Remove the shoes and springs from the brake panel. Check the ends of the shoes where they contact the brake cam and pivot post. Replace the shoes if there's visible wear.

7   Check the brake cam and pivot post for wear or damage. If necessary, make match marks on the cam and cam lever, then remove the pinch bolt, lever, wear indicator pointer, seal and cam **(see illustration 13.6c)**.

8   Check the drum (inside the wheel) for wear or damage, such as scoring, cracks or hard spots. Measure the diameter at several points with a brake drum micrometer **(see illustration)** or have this done by a Yamaha dealer. If the measurements are uneven (indicating the drum is out of round) or if there are scratches deep enough to snag a fingernail, have the drum turned (skimmed) by a dealer to correct the surface. If the drum has to be turned (skimmed) beyond the wear limit to correct defects, replace it.

9   Check the brake cam for looseness in the brake panel hole. If it's loose, replace the cam or brake panel, whichever is worn.

Installation

10   Apply high temperature brake grease to the ends of the springs, the cam and the anchor pin.

11   Hook the springs to the shoes. Position the shoes in a V on the

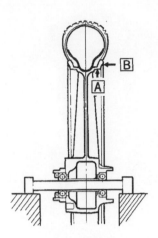

10.2 Check the wheel for out-of-round (A) and lateral movement (B)

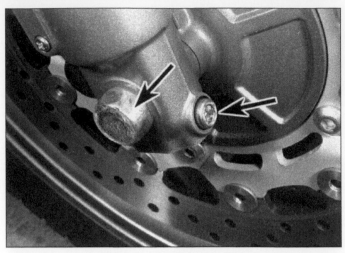

12.4a On some models, the axle (left arrow) is secured by a pinch bolt that threads into the fork leg (right arrow)

brake panel, then fold them down into position. Make sure the ends of the shoes fit correctly in the cam and on the anchor pin.

12   The remainder of installation is the reverse of the removal steps.

13   Check the brake pedal position and free play (see Chapter 1)and adjust them if necessary. Check the operation of the brakes carefully before riding the motorcycle.

10   Wheels - inspection and repair

Refer to illustration 10.2

1   Place the motorcycle on the centerstand, then clean the wheels thoroughly to remove mud and dirt that may interfere with the inspection procedure or mask defects. Make a general check of the wheels and tires as described in Chapter 1.

2   With the motorcycle on the centerstand (if equipped) or supported securely upright and the wheel in the air, attach a dial indicator to the fork slider or the swingarm and position the stem against the side of the rim **(see illustration)**. Spin the wheel slowly and check the side-to-side (axial) runout of the rim, then compare your readings with the value listed in this chapter's Specifications. In order to accurately check radial runout with the dial indicator, the wheel would have to be removed from the machine and the tire removed from the wheel. With the axle clamped in a vise, the wheel can be rotated to check the runout.

3   An easier, though slightly less accurate, method is to attach a stiff wire pointer to the fork slider or the swingarm and position the end a fraction of an inch from the wheel (where the wheel and tire join). If the wheel is true, the distance from the pointer to the rim will be constant as the wheel is rotated. Repeat the procedure to check the runout of the rear wheel. **Note:** *If wheel runout is excessive, refer to Section 14 and check the wheel bearings very carefully before replacing the wheel.*

4   The wheels should also be visually inspected for cracks, flat spots on the rim and other damage. Since tubeless tires are fitted, look very closely for dents in the area where the tire bead contacts the rim. Dents in this area may prevent complete sealing of the tire against the rim, which leads to deflation of the tire over a period of time.

5   If damage is evident, or if runout in either direction is excessive, the wheel will have to be replaced with a new one. Never attempt to repair a damaged cast aluminum wheel.

11   Wheels - alignment check

1   Misalignment of the wheels, which may be due to a cocked rear wheel or a bent frame or triple clamps, can cause strange and possibly serious handling problems. If the frame or triple clamps are at fault, repair by a frame specialist or replacement with new parts are the only alternatives.

12.4b On other models, the axle is secured by a pinch bolt and nut

2   To check the alignment you will need an assistant, a length of string or a perfectly straight piece of wood and a ruler graduated in 1/64 inch increments. A plumb bob or other suitable weight will also be required.

3   Place the motorcycle on the centerstand, then measure the width of both tires at their widest points. Subtract the smaller measurement from the larger measurement, then divide the difference by two. The result is the amount of offset that should exist between the front and rear tires on both sides.

4   If a string is used, have your assistant hold one end of it about half way between the floor and the rear axle, touching the rear sidewall of the tire.

5   Run the other end of the string forward and pull it tight so that it is roughly parallel to the floor. Slowly bring the string into contact with the front sidewall of the rear tire, then turn the front wheel until it is parallel with the string. Measure the distance from the front tire sidewall to the string.

6   Repeat the procedure on the other side of the motorcycle. The distance from the front tire sidewall to the string should be equal on both sides.

7   As was previously pointed out, a perfectly straight length of wood may be substituted for the string. The procedure is the same.

8   If the distance between the string and tire is greater on one side, or if the rear wheel appears to be cocked, refer to Chapter 5, *Swingarm bearings - check*, and make sure the swingarm is tight.

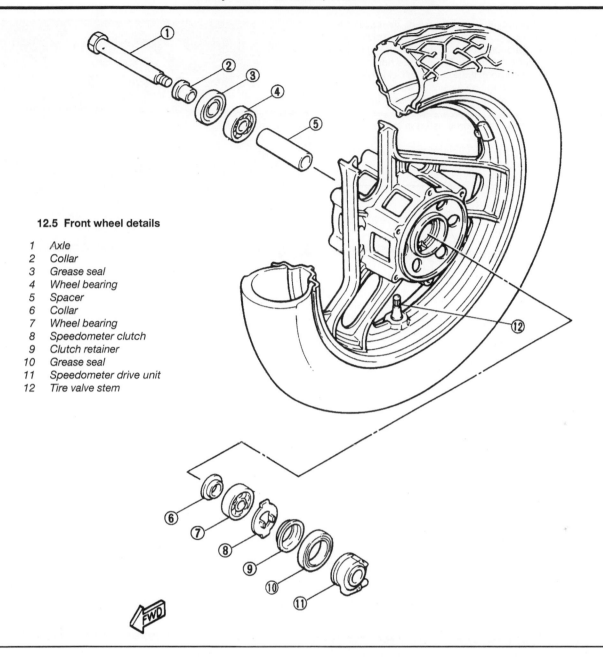

12.5 Front wheel details

1   Axle
2   Collar
3   Grease seal
4   Wheel bearing
5   Spacer
6   Collar
7   Wheel bearing
8   Speedometer clutch
9   Clutch retainer
10   Grease seal
11   Speedometer drive unit
12   Tire valve stem

9   If the front-to-back alignment is correct, the wheels still may be out of alignment vertically.

10   Using the plumb bob, or other suitable weight, and a length of string, check the rear wheel to make sure it is vertical. To do this, hold the string against the tire upper sidewall and allow the weight to settle just off the floor. When the string touches both the upper and lower tire sidewalls and is perfectly straight, the wheel is vertical. If it isn't, place thin spacers under one leg of the centerstand.

11   Once the rear wheel is vertical, check the front wheel in the same manner. If both wheels are not perfectly vertical, the frame and/or major suspension components are bent.

12   Front wheel - removal and installation

Refer to illustrations 12.4a, 12.4b, 12.5 and 12.7

Removal

1   Place the motorcycle on the centerstand (if equipped) or support it securely upright, then raise the front wheel off the ground by placing

a floor jack, with a wood block on the jack head, under the engine.

2   Disconnect the speedometer cable from the drive unit (see Chapter 8).

3   Remove both brake calipers and support the calipers with a piece of wire. don't disconnect the brake hoses from the calipers.

4   Remove the axle clamp bolt (and nut if equipped) **(see illustrations)** and unscrew the axle.

5   Support the wheel, then pull out the axle **(see illustration)** and carefully lower the wheel. **Caution:** *don't lay the wheel down and allow it to rest on one of the discs - the disc could become warped. Set the wheel on wood blocks so the disc doesn't support the weight of the wheel. Roll the axle on a flat surface such as a piece of plate glass. If it's bent at all, replace it. If the axle is corroded, remove the corrosion with fine emery cloth.* **Note:** *Do not operate the front brake lever with the wheel removed.*

6   Check the condition of the wheel bearings (see Section 14).

Installation

7   Installation is the reverse of removal. Apply a thin coat of grease to the seal lip, then slide the axle into the hub. Position the speedometer

drive unit in the left side of the hub (if it was removed), then slide the wheel into place. Make sure the lugs in the speedometer drive clutch line up with the notches in the speedometer drive unit. Make sure the protrusion on the inner side of the left fork fits into the notch in the speedometer drive unit **(see illustration)**. If the disc won't slide between the brake pads, carefully pry them apart with a piece of wood.

8　　Thread the front wheel axle into the fork leg and tighten it securely.

9　　Tighten the axle pinch bolt to the torque listed in this chapter's Specifications.

10　Apply the front brake, pump the forks up and down several times and check for binding and proper brake operation.

12.7 Align the lugs on the speedometer clutch with the slots in the drive unit (arrows)

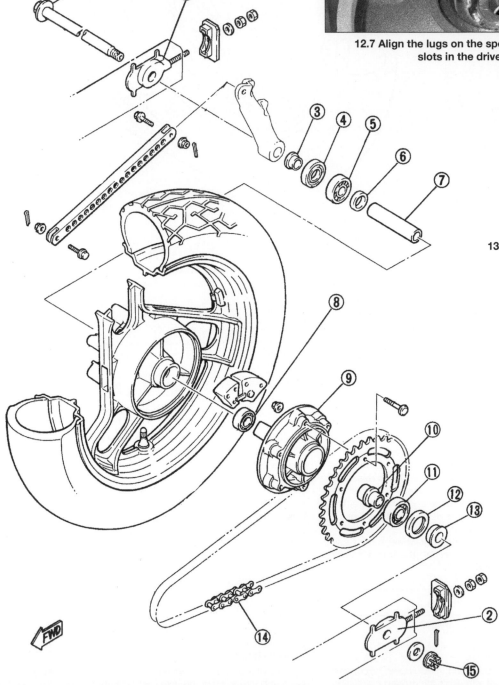

13.6a Rear wheel details - FZ600 models

1　Axle
2　Chain adjuster
3　Collar
4　Grease seal
5　Wheel bearing
6　Spacer flange
7　Spacer
8　Wheel bearing
9　Rear wheel coupling
10　Coupling collar
11　Coupling bearing
12　Grease seal
13　Collar
14　Drive chain
15　Axle nut

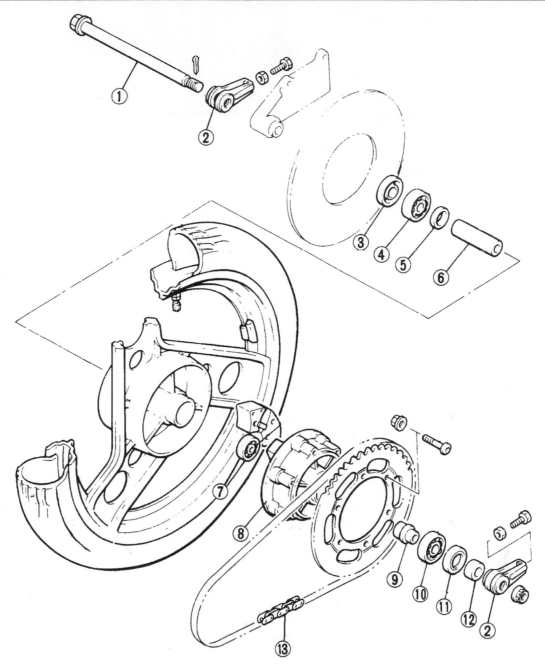

13.6b Rear wheel details - FJ600 and XJ600 models

1	Axle	6	Spacer	10	Wheel bearing
2	Chain adjuster	7	Wheel bearing	11	Grease seal
3	Grease seal	8	Coupling	12	Collar
4	Wheel bearing	9	Coupling collar	13	Drive chain
5	Spacer flange				

13 Rear wheel - removal and installation

Refer to illustrations 13.6a, 13.6b, 13.6c

Removal

1   Set the bike on its centerstand (if equipped) or prop it securely upright with the rear wheel off the ground.
2   Remove the chain guard (see Chapter 1 if necessary).
3   Remove the cotter pin from the axle nut and loosen the nut (see the

chain adjustment section of Chapter 1).
4   Loosen the chain adjusting bolt locknuts and fully loosen both adjusters.
5   Push the rear wheel as far forward as possible. Lift the top of the chain up off the rear sprocket and pull it to the left while rotating the wheel backwards. This will disengage the chain from the sprocket. **Warning**: *Don't let your fingers slip between the chain and the sprocket.*
6   Unscrew the axle nut **(see illustrations)**.

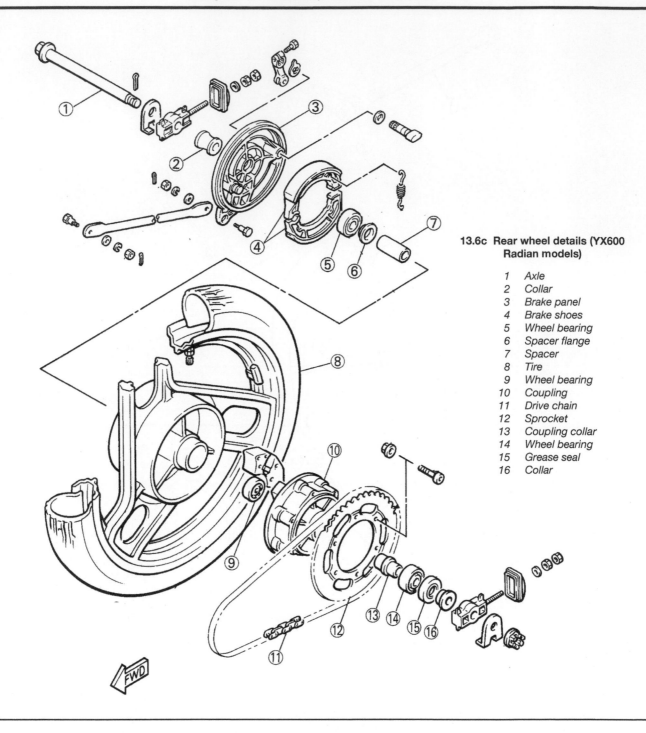

13.6c Rear wheel details (YX600 Radian models)

1   Axle
2   Collar
3   Brake panel
4   Brake shoes
5   Wheel bearing
6   Spacer flange
7   Spacer
8   Tire
9   Wheel bearing
10   Coupling
11   Drive chain
12   Sprocket
13   Coupling collar
14   Wheel bearing
15   Grease seal
16   Collar

7   If you're working on a YX600 Radian, remove the tension rod from the brake assembly **(see illustration)**. Mark the position of the brake operating lever on the cam, then remove the pinch bolt and slide the lever off **(see illustration).**

8   Support the wheel and slide the axle out. Lower the wheel and remove it from the swingarm, being careful not to lose the spacers on either side of the hub **(see illustration)**. If you're working on a disc brake model, slide the disc out from between the brake pads. **Caution:** *Don't lay the wheel down and allow it to rest on the disc or the sprocket - they could become warped. Set the wheel on wood blocks so that the disc or the sprocket Doesn't support the weight of the wheel. Don't operate the brake pedal with the wheel removed.*

9   Before installing the wheel, check the axle for straightness by rolling it on a flat surface such as a piece of plate glass (if the axle is

corroded, first remove the corrosion with fine emery cloth). If the axle is bent at all, replace it.

10   Check the condition of the wheel bearings (see Section 14).

Installation

11   Apply a thin coat of grease to the seal lips, then slide the spacers into their proper positions on the sides of the hub.

12   Slide the wheel into place, making sure the brake disc (if equipped) slides between the brake pads. If it Doesn't, spread the pads apart with a piece of wood.

13   Pull the chain up over the sprocket, raise the wheel and install the axle and axle nut. Don't tighten the axle nut yet.

14   Adjust the chain slack (see Chapter 1) and tighten the adjuster locknuts.

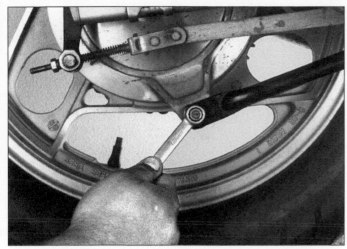

13.7a Remove the nut and bolt and detach the torque link from the brake panel

13.7b Remove the pinch bolt and slide the brake lever off the cam

13.8 Pull the rear wheel out, taking care not to lose the collar on each side of the wheel

14.10 Install the bearing with a drive (shown) or a socket the same diameter as the bearing outer race

15   Tighten the axle nut to the torque listed in this chapter's Specifications. Install a new cotter pin, tightening the axle nut an additional amount, if necessary, to align the hole in the axle with the castellations on the nut. Be sure to bend the cotter pin correctly (see *Drive chain and sprockets - check, adjustment and lubrication* in Chapter 1).

16   On drum brake models, install the tension rod and brake lever.

17   Check the operation of the brakes carefully before riding the motorcycle.

14   Wheel bearings - inspection and maintenance

Refer to illustration 14.10

1   Remove the wheel. See Section 12 (front wheel) or Section 13 (rear wheel).

2   Set the wheel down so it doesn't rest on the brake disc (if equipped) or sprocket (rear wheel).

Front wheel bearings

3   From the right side of the wheel, lift out the collar **(see illustration 12.5).**

4   From the left side of the wheel, lift out the speedometer drive unit.

5   From the left side of the wheel, pry out the grease seal, then lift out the speedometer clutch retainer and speedometer clutch.

6   Using a metal rod (preferably a brass drift punch) inserted through the center of the hub bearing, tap evenly around the inner race of the opposite bearing to drive it from the hub. The bearing spacer will also come out.

7   Lay the wheel on its other side and remove the remaining bearing using the same technique.

8   Clean the bearings with a high flash-point solvent (one which Won't leave any residue) and blow them dry with compressed air (Don't let the bearing spin as you dry them). Apply a few drops of oil to the bearing. Hold the outer race of the bearing and rotate the inner race - if the bearing Doesn't turn smoothly, has rough spots or is noisy, replace it with a new assembly.

9   If the bearing checks out okay and will be reused, wash it in solvent once again and dry it, then pack the bearing with high-quality bearing grease.

10   Thoroughly clean the hub area of the wheel. Install the bearing into the recess in the hub, with the marked or shielded side facing out. Using a bearing driver or a socket large enough to contact the outer race of the bearing, drive it in **(see illustration)** until It's completely seated.

11   Turn the wheel over and install the bearing spacer and bearing, driving the bearing into place as described in Step 9. Install the speedometer clutch and retainer on the left side of the wheel.

12   Install new grease seals, using a seal driver, large socket or a flat piece of wood to drive them into place.

14.15 Lift the collar out

14.20 Drive the wheel bearing out with a brass punch

13   Install the speedometer drive unit, making sure the lugs in the speedometer clutch align with the notches in the gear **(see illustration 12.7).**

14   Clean off all grease from the brake disc(s) using acetone or brake system cleaner. Install the wheel.

Rear coupling bearing

Refer to illustration 14.15

15   Lift the collar from the coupling on the sprocket side of the wheel **(see illustration)**. Pry out the grease seal.

16   Lift the sprocket out of the damper assembly and turn it over. Drive out the coupling bearing with a bearing driver or socket **(see illustrations 13.6a, 13.6b or 13.6c)**.

17   Hold the coupling bearing by the inner race and spin the outer race. If It's rough, loose or noisy, replace the bearing assembly.

18   Drive the bearing into the coupling with a bearing driver or socket.

Rear wheel bearings

Refer to illustration 14.20

19   If you haven't already done so, remove the collars from both sides of the wheel **(see illustrations 13.6a, 13.6b or 13.6c)**.

20   Insert a drift through one bearing and tap out the bearing on the opposite side **(see illustration)**. Remove the spacer and spacer flange.

21   Turn the wheel over and tap out the bearing on the opposite side.

22   Perform Steps 9 and 10 above to inspect the bearings.

23   Install the bearings with their sealed sides facing out. Be sure to install the spacer and spacer flange in the correct order and facing in the proper direction **(see illustrations 13.6a, 13.6b or 13.6c)**. Use a bearing driver or socket the same diameter as the outer race of the bearing.

24   Install the grease seals and collars.

25   Install the coupling to the wheel, making sure the coupling collar is in place. **Caution:** *If the coupling collar is left out, the rear wheel bearings will be damaged when the axle nut is tightened.*

15   Tubeless tires - general information

1   Tubeless tires are used as standard equipment on this motorcycle. They are generally safer than tube-type tires but if problems do occur they require special repair techniques.

2   The force required to break the seal between the rim and the bead of the tire is substantial, and is usually beyond the capabilities of an individual working with normal tire irons.

3   Also, repair of the punctured tire and replacement on the wheel rim requires special tools, skills and experience that the average do-it-yourselfer lacks.

4   For these reasons, if a puncture or flat occurs with a tubeless tire, the wheel should be removed from the motorcycle and taken to a dealer service department or a motorcycle repair shop for repair or replacement of the tire. The accompanying color illustrations can be used to replace a tubeless tire in an emergency.

TIRE CHANGING SEQUENCE - TUBELESS TIRES

1

Deflate tire. After releasing beads, push tire bead into well of rim at point opposite valve. Insert lever next to valve and work bead over edge of rim.

2

Use two levers to work bead over edge of rim. Note use of rim protectors.

3

When first bead is clear, remove tire as shown.

4

Before installing, ensure that tire is suitable for wheel. Take note of any sidewall markings such as direction of rotation arrows.

5

Work first bead over the rim flange.

6

Use a tire lever to work the second bead over rim flange.

Notes

Chapter 7
Fairings and bodywork

Contents

1 General information

This Chapter covers the procedures necessary to remove and install the fairings and other body parts for service access. Since many service and repair operations on these motorcycles require removal of the fairings or other body parts, the procedures are grouped here and referred to from other Chapters.

In the event of damage to the fairings or other body part, it is usually necessary to remove the broken component and replace it with a new (or used) one. The material that the fairings are composed of doesn't lend itself to conventional repair techniques. There are, however, some shops that specialize in "plastic welding," so it would be advantageous to check around first before throwing the damaged part away.

2 Seat - removal and installation

XJ600 and FJ600 models
Refer to illustrations 2.1 and 2.2
1   Turn the seat lock counterclockwise (anti-clockwise) with the key and pull down the latches **(see illustration).**
2   Pull the seat back and up to disengage its mounting tab **(see illustration).** Lift the seat out.

FZ600 models
3   Turn the seat lock clockwise with the key. Lift the passenger seat and disengage its lobes from the frame.
4   Remove the two bolts at the rear of the rider seat and lift it out.

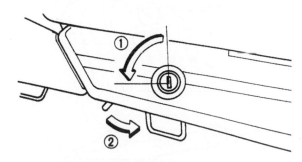

2.1 Turn the lock (1) counterclockwise (anti-clockwise) and pull down on the latches (2)

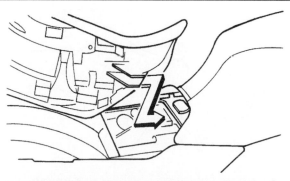

2.2 Disengage the tab at the front of the seat

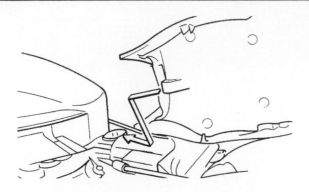

2.5 Disengage the tab at the front of the seat

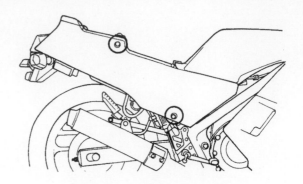

3.4a Remove the mounting bolts . . .

YX600 Radian models

Refer to illustration 2.5
5   Turn the seat lock clockwise with the key. Pull the seat up and back to disengage its mounting tab **(see illustration).**

All models
6   Installation is the reverse of the removal steps.

3   Side covers - removal and installation

1   Set the bike on its centerstand (if equipped) or prop it securely upright. Disengaging the side covers on some models requires gentle tugging which could tip the bike over if it isn't securely supported.

FJ/XJ600 models
2   Carefully pull the side cover outward at top front, bottom center and top rear to disengage it from the grommets.

FZ600 models
Refer to illustrations 3.4a and 3.4b
3   Remove the seat (see Section 2).
4   Remove the side cover bolts **(see illustration).** Pull the molded knobs out of the grommets and disengage the tab at the top rear **(see illustration).**

YX600 Radian models
Refer to illustration 3.5
5   Remove the screws at top and bottom of the side cover and disengage the side cover from the motorcycle **(see illustration).**

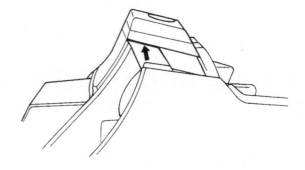

3.4b . . . and disengage the tab at the top rear

All models
6   Installation is the reverse of the removal steps.

4   Lower fairing (FJ/XJ600 models) - removal and installation

Refer to illustration 4.3
1   Support the motorcycle securely upright.
2   Support the fairing from below so it won't fall on the ground and be scratched.
3   Remove the fairing mounting screws **(see illustration).** Lower the fairing away from the motorcycle.
4   Installation is the reverse of the removal steps.

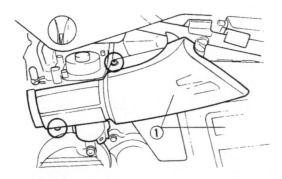

3.5 Remove the mounting screws to detach the side cover

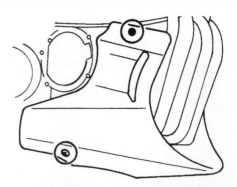

4.3 With the fairing supported from below, remove the mounting fasteners on each side

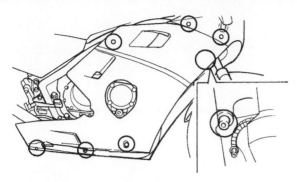

6.2 Remove the mounting fasteners along the sides of the fairing panel and two at the front near the fork legs

5 Upper fairing (FJ/XJ600 models) - removal and installation

1   Remove the headlight assembly (see Chapter 8).
2   Remove the fairing fasteners; there are two in the top of the headlight socket and two on each side of the fairing. Note the positions of the mounting grommets.
3   Installation is the reverse of the removal steps.

6 Lower and center fairings (FZ600 models) - removal and installation

Refer to illustration 6.2
1   Support the lower fairing panels from below so they won't fall on the ground and be scratched.
2   Remove the mounting fasteners along the sides of the fairing panels and inside the front near the fork legs **(see illustration)**. Lower the fairing panels and pull them out from under the motorcycle.
3   Installation is the reverse of removal.

7 Upper fairing (FZ600 models) - removal and installation

1   Remove the nuts inside the fairing and lift off the mirrors and their baseplate(s).
2   Remove the headlight covers (see Chapter 8).
3   Disconnect the turn signal electrical connectors, then remove the mounting nuts and take the turn signal stalks off the fairing.
4   Taking care not to let the headlight assemblies fall out of the fairing, remove two mounting fasteners on each side and lift the fairing off.
5   Installation is the reverse of removal. Tighten the mounting fasteners securely, but don't overtighten them and crack the plastic.

Notes

Chapter 8
Electrical system

Contents

Specifications

Battery

Capacity	12V, 12Ah
Specific gravity	See Chapter 1

Fuses

Main
All except YX600 Radian	30 amps
YX600 Radian	20 amps

Headlight
All except YX600 Radian	20 Amps
YX600 Radian	10 amps
Signal	10 amps
Ignition	10 amps

Fuel sender resistance

FJ600
Full	7 ohms +/- 5% at 20-degrees C (68-degrees F)
Empty	95 ohms +/- 7.5% at 20-degrees C (68-degrees F)

FZ600
Full	8.7 to 14.7 ohms at 20-degrees C (68-degrees F)
Empty	125 to 145 ohms at 20-degrees C (68-degrees F)

XJ600
Full	6.78 to 7.4 ohms at 20-degrees C (68-degrees F)
Empty	88 to 102 ohms at 20-degrees C (68-degrees F)

Starter motor

Starter commutator diameter
    Standard... 28 mm (1.1 inch)
    Minimum.. 27 mm (1.06 inch)
Starter brush length
    FJ600, FZ600 and YX600 Radian
        Standard .... 12 mm (0.47 inch)
        Minimum .... 5.0 mm (0.20 inch)
    XJ600
        Standard .... 12.5 mm (0.49 inch)
        Minimum .... 4.0 mm (0.16 inch)
Starter relay resistance
    FJ600 .... Not specified
    FZ600 .... 3.4 ohms at 20-degrees C (68-degrees F)
    XJ600 .... 3.9 to 4.7 ohms at 20-degrees C (68-degrees F)
    YX600 Radian.. 4.3 ohms at 20-degrees C (68-degrees F)

Alternator

Charging voltage .... 14 to 15 volts at 5000 rpm
Nominal output
    FJ600, FZ600, YX600 Radian .... 14 volts, 17 amps at 5000 rpm
    XJ600 .... 14 volts, 21 amps at 5000 rpm
No-load regulated voltage
    FJ600, FZ600, YX600 Radian .... 14.2 to 14.8 volts
    XJ600 .... 14.3 to 15.3 volts
Stator coil resistance
    FJ600, FZ600, YX600 Radian .... 0.5 to 0.6 ohms at 20-degrees C (68-degrees F)
    XJ600 .... 0.31 to 0.37 ohms at 20-degrees C (68-degrees F)
Field coil resistance (early models).. 2.7 to 3.3 ohms at 20-degrees C (68-degrees F)
Brush length (early models)
    Standard.. 17 mm (0.669 inch)
    Minimum.. 10 mm (0.394 inch)

Bulbs

Headlight 60/55W
Stop/taillight
    US 8/27W
    UK 5/21W
Parking light (UK only) 3.4W
Turn signal lights
    US 27W
    UK 21W
Meter lights
    All except YX600 Radian.. 3.4W
    YX600 Radian.. 3W
Indicator lights 3.4W
License plate light
    XJ/FJ600 .... not specified
    FZ600, YX600 Radian 3.8W

Torque specifications

Oil level sender bolts .... 10 Nm (7.2 ft-lbs)
Alternator (early models with brushes)
    Rotor bolt .... 35 Nm (25 ft-lbs)
    Brush assembly screws .... 8 Nm (5.8 ft-lbs)
Alternator (later models without brushes)
    Rotor bolt .... 80 Nm (58 ft-lbs)
    Stator coil Allen bolts .... 10 Nm (7.2 ft-lbs)*
    Pickup coil screws .... 5 Nm (3.6 ft-lbs)
    Stator coil wiring harness screws .... 7 Nm (5.1 ft-lbs)
    Cover Allen bolts .... 10 Nm (7.2 ft-lbs)
Starter mounting bolts.. 10 Nm (7.2 ft-lbs)

*Apply non-permanent thread locking agent to the threads.

1 General information

The machines covered by this manual are equipped with a 12-volt electrical system.

The regulator/rectifier maintains the charging system output within the specified range to prevent overcharging and converts the AC (alternating current) output of the alternator to DC (direct current) to power the lights and other components and to charge the battery.

The alternator on early models uses an electromagnetic field, which requires a brush assembly. Later models use permanent magnets mounted in the rotor; this design requires no brushes.

An electric starter mounted to the engine case behind the cylinders is standard equipment. The starting system includes the motor, the battery, the solenoid and the various wires and switches. On models equipped with a sidestand switch and clutch switch, if the engine kill switch and the main key switch are both in the On position, the circuit relay allows the starter motor to operate only if the transmission is in Neutral (Neutral switch on) or the clutch lever is pulled to the handlebar (clutch switch on) and the sidestand is up (sidestand switch on).

Note: *Keep in mind that electrical parts, once purchased, can't be returned. To avoid unnecessary expense, make very sure the faulty component has been positively identified before buying a replacement part.*

2 Electrical troubleshooting

A typical electrical circuit consists of an electrical component, the switches, relays, etc. related to that component and the wiring and connectors that hook the component to both the battery and the frame. To aid in locating a problem in any electrical circuit, complete wiring diagrams of each model are included at the end of this Chapter.

Before tackling any troublesome electrical circuit, first study the appropriate diagrams thoroughly to get a complete picture of what makes up that individual circuit. Trouble spots, for instance, can often be narrowed down by noting if other components related to that circuit are operating properly or not. If several components or circuits fail at one time, chances are the fault lies in the fuse or ground (earth) connection, as several circuits often are routed through the same fuse and ground (earth) connections.

Electrical problems often stem from simple causes, such as loose or corroded connections or a blown fuse. Prior to any electrical troubleshooting, always visually check the condition of the fuse, wires and connections in the problem circuit. Intermittent failures can be especially frustrating, since you can't always duplicate the failure when it's convenient to test. In such situations, a good practice is to clean all connections in the affected circuit, whether or not they appear to be good. All of the connections and wires should also be wiggled to check for looseness which can cause intermittent failure.

If testing instruments are going to be utilized, use the diagrams to plan where you will make the necessary connections in order to accurately pinpoint the trouble spot.

The basic tools needed for electrical troubleshooting include a test light or voltmeter, a continuity tester (which includes a bulb, battery and set of test leads) and a jumper wire, preferably with a circuit breaker incorporated, which can be used to bypass electrical components. Specific checks described later in this Chapter require an ohmmeter.

Voltage checks should be performed if a circuit is not functioning properly. Connect one lead of a test light or voltmeter to either the negative battery terminal or a known good ground (earth). Connect the other lead to a connector in the circuit being tested, preferably nearest to the battery or fuse. If the bulb lights, voltage is reaching that point, which means the part of the circuit between that connector and the battery is problem-free. Continue checking the remainder of the circuit in the same manner. When you reach a point where no voltage is present, the problem lies between there and the last good test point.

Most of the time the problem is due to a loose connection. Keep in mind that some circuits only receive voltage when the ignition key is in the On position.

One method of finding short circuits is to remove the fuse and connect a test light or voltmeter in its place to the fuse terminals. There should be no load in the circuit (it should be switched off). Move the wiring harness from side-to-side while watching the test light. If the bulb lights, there is a short to ground (earth) somewhere in that area, probably where insulation has rubbed off a wire. The same test can be performed on other components in the circuit, including the switch.

A ground (earth) check should be done to see if a component is grounded (earthed) properly. Disconnect the battery and connect one lead of a self-powered test light (continuity tester) to a known good ground (earth). Connect the other lead to the wire or ground (earth) connection being tested. If the bulb lights, the ground (earth) is good. If the bulb does not light, the ground (earth) is not good.

A continuity check is performed to see if a circuit, section of circuit or individual component is capable of passing electricity through it. Disconnect the battery and connect one lead of a self-powered test light (continuity tester) to one end of the circuit being tested and the other lead to the other end of the circuit. If the bulb lights, there is continuity, which means the circuit is passing electricity through it properly. Switches can be checked in the same way.

Remember that all electrical circuits are designed to conduct electricity from the battery, through the wires, switches, relays, etc. to the electrical component (light bulb, motor, etc.). From there it is directed to the frame (ground (earth)) where it is passed back to the battery. Electrical problems are basically an interruption in the flow of electricity from the battery or back to it.

3 Battery - inspection and maintenance

1   Most battery damage is caused by heat, vibration, and/or low electrolyte levels, so keep the battery securely mounted, check the electrolyte level frequently and make sure the charging system is functioning properly.
2   Refer to Chapter 1 for electrolyte level and specific gravity checking procedures.
3   Check around the base inside of the battery for sediment, which is the result of sulfation caused by low electrolyte levels. These deposits will cause internal short circuits, which can quickly discharge the battery. Look for cracks in the case and replace the battery if either of these conditions is found.
4   Check the battery terminals and cable ends for tightness and corrosion. If corrosion is evident, remove the cables from the battery and clean the terminals and cable ends with a wire brush or knife and emery paper. Reconnect the cables and apply a thin coat of petroleum jelly to the connections to slow further corrosion.
5   The battery case should be kept clean to prevent current leakage, which can discharge the battery over a period of time (especially when it sits unused). Wash the outside of the case with a solution of baking soda and water. Do not get any baking soda solution in the battery cells. Rinse the battery thoroughly, then dry it.
6   If acid has been spilled on the frame or battery box, neutralize it with the baking soda and water solution, dry it thoroughly, then touch up any damaged paint. Make sure the battery vent tube (if equipped) is directed away from the frame and is not kinked or pinched.
7   If the motorcycle sits unused for long periods of time, disconnect the cables from the battery terminals. Refer to Section 4 and charge the battery approximately once every month.

4 Battery - charging

1   If the machine sits idle for extended periods or if the charging system malfunctions, the battery can be charged from an external source.

5.1a The fuse block is located under the seat (arrow) . . .

5.1b Remove the cover to inspect the fuses

2   To properly charge the battery, you will need a charger of the correct rating, a hydrometer, a clean rag and a syringe for adding distilled water to the battery cells.

3   The maximum charging rate for any battery is 1/10 of the rated amp/hour capacity. As an example, the maximum charging rate for the 14 amp/hour battery would be 1.4 amps. If the battery is charged at a higher rate, it could be damaged.

4   Do not allow the battery to be subjected to a so-called quick charge (high rate of charge over a short period of time) unless you are prepared to buy a new battery.

5   When charging the battery, always remove it from the machine and be sure to check the electrolyte level before hooking up the charger. Add distilled water <u>only</u> to any cells that are low.

6   Loosen the cell caps, hook up the battery charger leads (red to positive, black to negative), cover the top of the battery with a clean rag, then, and only then, plug in the battery charger. **Warning**: *Remember, the gas escaping from a charging battery is explosive, so keep open flames and sparks well away from the area. Also, the electrolyte is extremely corrosive and will damage anything it comes in contact with.*

7   Allow the battery to charge until the specific gravity is as specified (refer to Chapter 1 for specific gravity checking procedures). The charger must be unplugged and disconnected from the battery when making specific gravity checks. If the battery overheats or gases excessively, the charging rate is too high. Either disconnect the charger or lower the charging rate to prevent damage to the battery.

8   If one or more of the cells does not show an increase in specific gravity after a long slow charge, or if the battery as a whole does not seem to want to take a charge, it is time for a new battery.

9   When the battery is fully charged, unplug the charger first, then disconnect the leads from the battery. Install the cell caps and wipe any electrolyte off the outside of the battery case.

5   Fuses - check and replacement

Refer to illustrations 5.1a and 5.1b

1   All models have one fuse block located under the seat **(see illustrations)**. They include the main, headlight, signal and ignition fuses, as well as spares. Fuse ratings are listed in this Chapter's Specifications.

2   If you have a test light, all of the fuses can be checked without removing them. Turn the ignition to the On position, connect one end of the test light to a good ground (earth), then probe each terminal on top of the fuse. If the fuse is good, there will be voltage available at both terminals. If the fuse is blown, there will only be voltage present at one of the terminals.

3   The fuses can also be tested with an ohmmeter or self-powered test light. Remove the fuse and connect the tester to the ends of the fuse. If the ohmmeter shows continuity or the test lamp lights, the fuse is good. If the ohmmeter shows infinite resistance or the test lamp stays out, the fuse is blown.

4   The fuses can be checked visually. A blown fuse is easily identified by a break in the element.

5   If a fuse blows, be sure to check the wiring harnesses very carefully for evidence of a short-circuit. Look for bare wires and chafed, melted or burned insulation. If a fuse is replaced before the cause is located, the new fuse will blow immediately.

6   Never, under any circumstances, use a higher rated fuse or bridge the fuse block terminals, as damage to the electrical system could result.

7   Occasionally a fuse will blow or cause an open-circuit for no obvious reason. Corrosion of the fuse ends and fuse block terminals may occur and cause poor fuse contact. If this happens, remove the corrosion with a wire brush or emery paper, then spray the fuse end and terminals with electrical contact cleaner.

6   Lighting system - check

1   The battery provides power for operation of the headlight, taillight, brake light, license plate light and instrument cluster lights. If none of the lights operate, always check battery voltage before proceeding. Low battery voltage indicates either a faulty battery, low battery electrolyte level or a defective charging system. Refer to Chapter 1 for battery checks and Sections 27 and 28 for charging system tests. Also, check the condition of the fuses and replace any blown fuses with new ones.

Headlight

2   If the headlight is out when the engine is running (US models) or it won't switch on (UK models), check the fuse first with the key On (see Section 5), then unplug the electrical connector for the headlight and use jumper wires to connect the bulb directly to the battery terminals. If the light comes on, the problem lies in the wiring or one of the switches in the circuit. Refer to Section 17 for the switch testing procedures, and also the wiring diagrams at the end of this Chapter.

Taillight/license plate light

3   If the taillight fails to work, check the bulbs and the bulb terminals first, then check for battery voltage at the taillight electrical connector. If voltage is present, check the ground (earth) circuit for an open or poor connection.

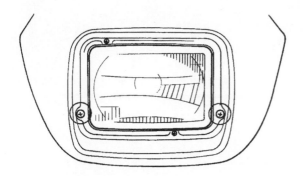

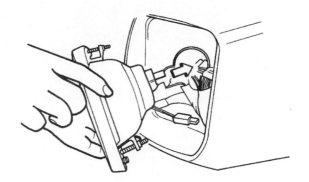

7.1a The headlight retaining screws on FJ/XJ600 models are located on either side of the headlight assembly (on YX600 Radian models they're on the lower corners of the headlight assembly); remove the screws . . .

7.1b . . . and pull the assembly out

7.2 Disconnect the electrical connector (arrow) and pull off the dust cover (note the Top marking on the cover)

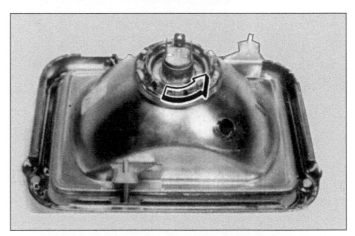

7.3a Rotate the retainer ring counterclockwise to free the bulb . . .

4   If no voltage is indicated, check the wiring between the taillight and the ignition switch, then check the switch. On UK models, check the lighting switch as well.

Brake light

5   See Section 11 for the brake light switch checking procedure.

Neutral indicator light

6   If the neutral light fails to operate when the transmission is in Neutral, check the fuses and the bulb (see Section 14 for bulb removal procedures). If the bulb and fuses are in good condition, check for battery voltage at the connector attached to the neutral switch on the left side of the engine. If battery voltage is present, refer to Section 19 for the neutral switch check and replacement procedures.
7   If no voltage is indicated, check the wiring between the switch and the bulb for open-circuits and poor connections.

Oil level warning light

8   See Section 15 for the oil level sender check.

---

7   Headlight bulb - replacement

Warning: *If the bulb has just burned out, allow it to cool. It will be hot enough to burn your fingers.*

7.3b . . . and pull the bulb out; be careful not to touch the bulb glass (1) with your fingers

FJ/XJ600 and YX600 Radian models
Refer to illustrations 7.1a, 7.1b, 7.2, 7.3a and 7.3b
1   Remove two screws and pull the headlight assembly out of the housing **(see illustrations)**.
2   Unplug the electrical connector from the headlight, then remove the dust cover **(see illustration)**.
3   Rotate the retainer counterclockwise (anti-clockwise) **(see illustration)** and remove the bulb from the holder **(see illustration)**.

7.4 On FZ600 models, pull the cover from the back of the headlight assembly

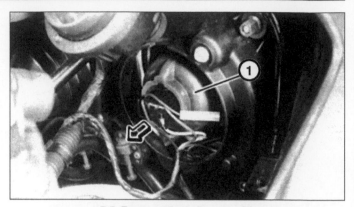

7.5 Remove the bulb cover (1)

7.6 Rotate the retainer ring counterclockwise to free the bulb . . .

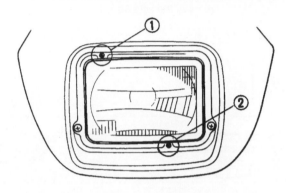

8.3 The adjuster screws on FJ/XJ600 models are accessible from the front

1   *Horizontal adjusting screw*   2   *Vertical adjusting screw*

FZ600 models

Refer to illustrations 7.4, 7.5 and 7.6

4   Reach inside the fairing and pull off the headlight cover **(see illustration)**.
5   Disconnect the electrical connector and remove the bulb cover **(see illustration)**.
6   Turn the retainer counterclockwise (anti-clockwise) and remove the holder **(see illustration)**. Remove the bulb from the holder.

All models

7   When installing the new bulb, reverse the removal procedure. Be sure not to touch the bulb's glass portion with your fingers - oil from your skin will cause the bulb to overheat and fail prematurely. If you do touch the bulb, wipe it off with a clean rag dampened with rubbing alcohol.

8   Headlight aim - check and adjustment

1   An improperly adjusted headlight may cause problems for oncoming traffic or provide poor, unsafe illumination of the road ahead. Before adjusting the headlight, be sure to consult with local traffic laws and regulations.
2   The headlight beam can be adjusted both vertically and horizontally. Before performing the adjustment, make sure the fuel tank is at least half full, and have an assistant sit on the seat.

FJ600 and XJ600 models

Refer to illustration 8.3

3   Vertical and horizontal adjustments are made by turning the screws at the front of the headlight assembly **(see illustration)**.

8.4 The adjusting screws on FZ600 models are accessible from behind

1   *Horizontal adjusting screw*   2   *Vertical adjusting screw*

FZ600 models

Refer to illustration 8.4

4   To make vertical and horizontal adjustments, reach inside the fairing and turn the adjusting screws **(see illustration)**.

YX600 Radian models

Refer to illustrations 8.5 and 8.6

5   Remove the trim cover **(see illustration)**.
6   To make vertical and horizontal adjustments, turn the adjusting screws **(see illustration)**.

8.5 To reach the adjusting screws on YX600 Radian models, remove the trim cover

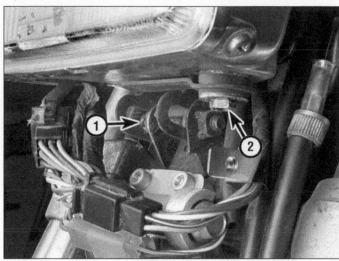

8.6 The horizontal adjusting screw (1) and vertical adjusting screw (2) are located below the headlight assembly

9.1 To replace a turn signal bulb, remove the screws and take off the lens

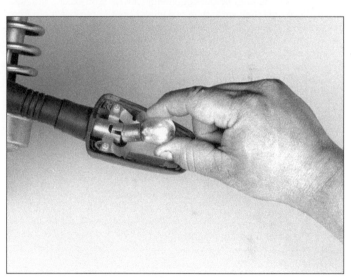

9.2 Press the bulb into its socket, turn counterclockwise (anti-clockwise) and pull it out

9 Turn signal and taillight bulbs - replacement

Turn signal bulbs

Refer to illustrations 9.1 and 9.2

1 On all models the turn signals are contained in stalks attached to the body panels or the frame. To replace a bulb, remove the lens securing screw(s) on the back of the assembly **(see illustration)**.

2 Push the bulb in and turn it counterclockwise (anti-clockwise) to remove it **(see illustration)**. Check the socket terminals for corrosion and clean them if necessary. Line up the pins on the new bulb with the slots in the socket, push in and turn the bulb clockwise until it locks in place. **Note**: *The pins on the bulb are offset so it can only be installed one way. It is a good idea to use a paper towel or dry cloth when handling the new bulb to prevent injury if the bulb should break and to increase bulb life.*

Tail/brake light bulbs

FJ/XJ600

3 Remove the lens securing screws and take off the lens.

4 Press the bulb into its socket and turn it counterclockwise (anti-clockwise) to disengage the pins, then pull it out. Check the socket terminals for corrosion and clean them if necessary. Line up the pins on the new bulb with the slots in the socket, push in and turn the bulb clockwise until it locks in place. **Note**: *The pins on the bulb are offset so it can only be installed one way. It is a good idea to use a paper towel or dry cloth when handling the new bulb to prevent injury if the bulb should break and to increase bulb life.*

5 Install the lens and tighten the screws securely, but not tight enough to crack the lens.

FZ600 and YX600 Radian models

Refer to illustrations 9.8a and 9.8b

6 Remove the seat (see Chapter 7).

9.8a To replace taillight bulbs on YX600 Radian models, reach inside the tail cowl, turn the bulb socket counterclockwise (anti-clockwise) and pull it out ...

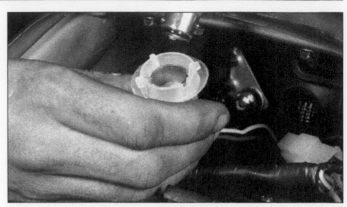

9.8b ... then push the bulb into its socket, turn counterclockwise (anti-clockwise) and pull it out

7   If you're working on an FZ600, remove four screws and remove the tail cowl.

8   The taillight can be removed by reaching into the space behind the seat. Turn the bulb holders counterclockwise (anti-clockwise) **(see illustration)** until they stop, then pull straight out to remove them from the taillight housing. The bulbs can be removed from the holders by pressing them in, turning them counterclockwise (anti-clockwise) and pulling straight out **(see illustration)**.

9   Check the socket terminals for corrosion and clean them if necessary. Line up the pins on the new bulb with the slots in the socket, push in and turn the bulb clockwise until it locks in place. **Note**: *The pins on the bulb are offset so it can only be installed one way. It is a good idea to use a paper towel or dry cloth when handling the new bulb to prevent injury if the bulb should break and to increase bulb life.*

10   Make sure the rubber gaskets (if equipped) are in place and in good condition, then line up the tabs on the holder with the slots in the housing and push the holder into the mounting hole. Turn it clockwise until it stops to lock it in place.

11   Reinstall all components removed for access.

10 Turn signal circuit - check

1   The battery provides power for operation of the signal lights, so if they do not operate, always check the battery voltage and specific gravity first. Low battery voltage indicates either a faulty battery, low electrolyte level or a defective charging system. Refer to Chapter 1 for battery checks and Sections 27 and 28 for charging system tests. Also, check the fuses (see Section 5) and the switch (see Section 18).

2   Most turn signal problems are the result of a burned out bulb or corroded socket. This is especially true when the turn signals function properly in one direction, but fail to flash in the other direction. Check the bulbs and the sockets (see Section 9).

FJ600, FZ600 and early Radian models

Refer to illustration 10.3

3   These models use a combined relay assembly that includes the flasher relay **(see illustration)**.

4   If the bulbs and sockets check out okay, turn the ignition switch On and check for voltage at the brown wire terminal on the relay assembly (connect the voltmeter positive lead to the brown wire terminal and the negative lead to bare metal on the frame. If there's no voltage at the terminal, check the wiring from the signal fuse to the terminal for a break or bad connection.

5   If there's voltage at the brown wire terminal, move the voltmeter positive lead to the brown/white wire terminal at the relay assembly and check for voltage again (the ignition switch should still be On). If there's no voltage, the flasher relay is probably defective; replace the relay assembly.

6   If there's voltage when there should in Steps 4 and 5 above, and if the turn signal switch tested OK, the most likely problem is a poor contact in a turn signal bulb socket. Check the sockets for corrosion or damage. Another possible cause is a worn contact at the base of the bulb. Try a known good bulb in the socket that won't work.

XJ600 and later YX600 Radian models

Refer to illustration 10.7

7   These models use a separate turn signal flasher **(see illustration)**.

8   With the ignition (main key) switch On, check for voltage at the brown wire terminal on the turn signal flasher (connect the voltmeter positive lead to the brown wire terminal and the negative lead to bare metal on the frame). If there's no voltage at the terminal, check the wiring from the ignition switch to the brown wire terminal for a break or bad connection.

9   If there's voltage at the brown wire terminal, move the voltmeter positive lead to brown/white wire terminal and check for voltage again (the ignition switch should still be On). If there's no voltage, the flasher unit is probably defective; replace it.

10   If the flasher is okay, check the wiring between the turn signal flasher and the turn signal light sockets (see the *wiring diagrams* at the end of this Chapter).

11 Brake light switches - check and replacement

Circuit check

1   Before checking any electrical circuit, check the fuses (see Section 5).

2   Using a test light connected to a good ground (earth), check for voltage at the brake light switch. If there's no voltage present, check the wire between the switch and the fuse block (see the *wiring diagrams* at the end of this Chapter).

3   If voltage is available, touch the probe of the test light to the other terminal of the switch, then pull the brake lever or depress the brake pedal - if the test light doesn't light up, replace the switch.

4   If the test light does light, check the wiring between the switch and the brake lights (see the *wiring diagrams* at the end of this Chapter).

Switch replacement

Front brake lever switch

Refer to illustration 11.5

5   Remove the mounting screw and unplug the electrical connector from the switch **(see illustration)**.

6   Detach the switch from the brake lever bracket/front master cylinder.

7   Installation is the reverse of the removal procedure. The brake lever switch isn't adjustable.

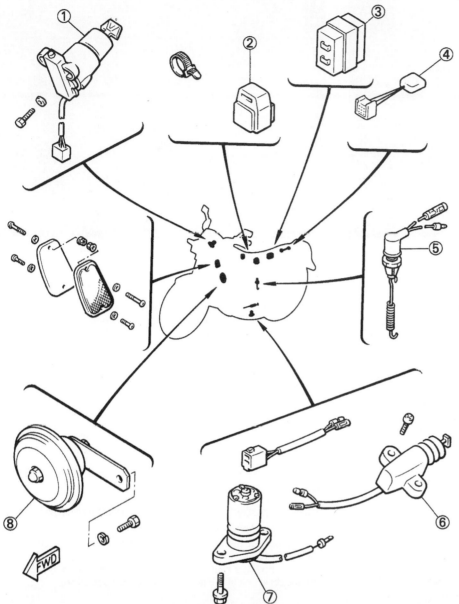

10.3 Electrical component locations (early YX600 Radian shown; others similar)

1   Ignition (main key) switch
2   Sidestand relay
3   Relay assembly
4   Diode assembly
5   Rear brake light switch
6   Sidestand switch
7   Oil level switch
8   Horn

10.7 The flasher relay on later YX600 Radian models is mounted on the right side of the air box

11.5 Disconnect the wires and remove the screw (arrow) to remove the front brake light switch

11.10 Disconnect the spring (arrow), unplug the electrical connector and rotate the plastic nut to remove the switch

12.4 Remove the screw and detach the instrument casing

Rear brake pedal switch

Refer to illustration 11.10

8   Unplug the electrical connector in the switch harness.

9   If necessary for access, remove the right footpeg or aluminum protective cover from the bike.

10   Disconnect the spring from the switch. Follow the wiring harness from the switch to the connector and disconnect it **(see illustration)**.

11   Hold the switch body to prevent it from turning and rotate the plastic adjusting nut until the switch threads separate from the nut. Pull the spring through the nut and lift the switch out.

12   Install the switch by reversing the removal procedure, then adjust the switch by following the procedure described in Chapter 1.

---

12   Instruments and speedometer cable - removal and installation

---

Instrument removal and installation

FJ600, FZ600 and XJ600 models

1   Remove the upper fairing (see Chapter 1).

2   Unscrew the knurled nut and disconnect the speedometer cable

from the speedometer.

3   Carefully pull the bulb sockets from the instrument being removed. Remove the mounting nuts and grommets and remove the cluster (FJ/XJ600) or individual gauges (FZ600).

YX600 Radian models

Refer to illustrations 12.4, 12.5, 12.6, 12.7a and 12.7b

4   Remove the Phillips screws used to retain the instrument casings **(see illustration)**.

5   Detach the speedometer cable from the speedometer **(see illustration)**.

6   Carefully pull the upper bulb socket from the instrument being removed, then remove the mounting nuts and lift the instrument out **(see illustration)**.

7   Pull out the remaining bulb socket(s) and lift out the instrument **(see illustrations)**.

Speedometer cable removal and installation

Refer to illustration 12.9

8   Disconnect the speedometer cable from the speedometer **(see illustration 12.5)**.

9   Note how it's routed, then disconnect the speedometer cable from the drive unit at the left front fork **(see illustration)**.

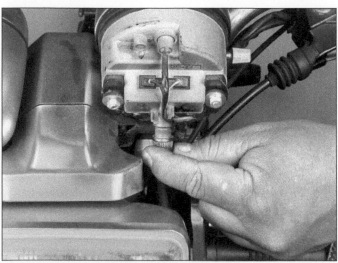

12.5 Unscrew the knurled nut and pull the speedometer cable out of the speedometer

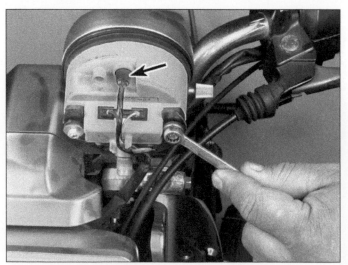

12.6 Pull out the bulb socket (arrow) and unscrew the mounting nuts

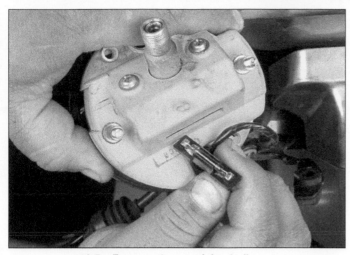

12.7a Remove the remaining bulbs . . .

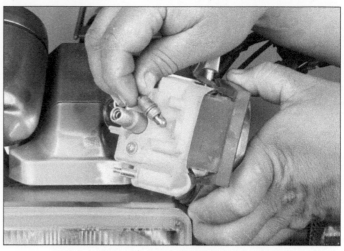

12.7b . . . from the speedometer or tachometer

10   Installation is the reverse of the removal procedure. Be sure the speedometer cable is routed so it doesn't cause the steering to bind or interfere with other components.

---

13   Meters and gauges - check

Fuel gauge (FJ600, FZ600 and XJ600)

1   To check the operation of the fuel gauge, remove the seat (see Chapter 7). Remove body panels as necessary for access and unplug the electrical connector from the fuel level sending unit.

2   Check for voltage with a voltmeter or test light at the green wire connector in the sending unit harness (the wiring harness side, not the wire that leads back to the fuel tank).

3   If there's voltage with the ignition switch On, test the sender (see Step 4 below). If not, check the wiring for breaks or bad connections.

4   If the sender tests okay, disconnect the fuel gauge wiring connector and check for voltage at the power supply terminal in the harness with the ignition switch On (refer to *wiring diagrams* at the end of this Chapter to locate the power terminal. If there's voltage, the gauge is probably defective. If there's no voltage, check the main and signal fuses, the ignition switch and the battery (Sections 5, 16 and 3).

Sender test

5   Remove the fuel tank (see Chapter 3). Remove the sender form the tank.

6   Connect an ohmmeter between the terminals of the sender electrical connector (the sender side, not the wiring harness side.

7   Move the sender float from the empty to the full positions and note the ohmmeter readings. If they're not within the range listed in this Chapter's Specifications, replace the sender.

Tachometer and speedometer

8   Special instruments are required to properly check the operation of these meters. Take the instrument cluster to a Yamaha dealer service department or other qualified repair shop for diagnosis.

---

14   Instrument and warning light bulbs - replacement

Refer to illustration 14.2

1   To replace instrument bulbs, remove the upper fairing or instrument housing as necessary for access. **Note:** *On FZ600 models, you should be able to replace instrument and warning light bulbs by reaching behind the instrument cluster and pulling the appropriate socket out of the instrument or warning light assembly.*

2   To replace warning light bulbs on YX600 Radian models,

12.9 Unscrew the knurled nut and pull the speedometer cable out of the drive unit at the fork leg

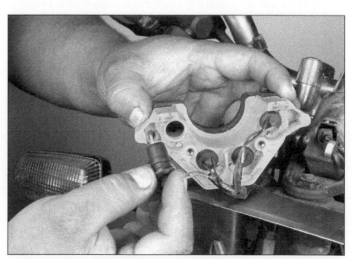

14.2 Pull the bulb socket out of the warning light assembly, then pull the bulb out of the socket

15.3 Remove the bolts and pull the oil level sender out of the oil pan; use a new O-ring for installation

SWITCH POSITION	WIRE COLOR		
	R	BR	L
ON	O—————O—————O
OFF	
P	O—————————————O

US MODELS

UK MODELS

	R	BR	L	L/R
ON	O——O		O——O
OFF				
P	O——————————————O

16.2 Ignition (main key) switch continuity diagram

remove the warning light assembly. Pull the appropriate rubber socket out of the back of the instrument or cluster **(see illustration 12.7b and the accompanying illustration)**, then pull the bulb out of the socket. If the socket contacts are dirty or corroded, they should be scraped clean and sprayed with electrical contact cleaner before new bulbs are installed.

3   Carefully push the new bulb into position, then push the socket into the cluster housing.

15 Oil level sender - removal, check and installation

Refer to illustration 15.3

Removal

1   Drain the engine oil (see Chapter 1).
2   The oil level sender is mounted in the bottom of the oil pan. Note how its wiring harness is routed, then unplug the electrical connector.
3   Remove the sender mounting bolts and remove the sender **(see illustration)**.

Check

4   Connect an ohmmeter between the terminals of the sender harness. With the sender in its normal installed position (flange and wiring harness at the bottom), the ohmmeter should indicate infinite resistance.
5   Turn the sender upside down. The ohmmeter should now read zero ohms.
6   If the ohmmeter doesn't give the correct indication in Step 4 or 5, replace the sender.

Installation

7   Installation is the reverse of the removal steps. Use a new O-ring and tighten the sender mounting bolts to the torque listed in this chapter's Specifications.

16 Ignition main (key) switch - check and replacement

Check

Refer to illustration 16.2

1   Follow the wiring harness from the ignition switch to the electrical connector. Remove fairing and instrument cluster components as

16.5 Remove the mounting bolts to detach the ignition (main key) switch

necessary for access and disconnect the connector.
2   Using an ohmmeter, check the continuity of the terminal pairs indicated in the accompanying table **(see illustration)**. Continuity should exist between the terminals connected by a solid line when the switch is in the indicated position.
3   If the switch fails any of the tests, replace it.

Replacement

Refer to illustration 16.5

4   Unplug the switch electrical connector if you haven't already done so.
5   The switch is held to the upper triple clamp with two bolts **(see illustration)**. Unlock the switch with the ignition key and remove the bolts. Detach the switch from the upper triple clamp.
6   If necessary, remove the Phillips screws and separate the switch from the bracket.
7   Attach the new switch to the bracket with the Phillips screws (if it was removed). Tighten the screws securely. Hold the new switch in position and install the bolts.
8   The remainder of installation is the reverse of the removal procedure.

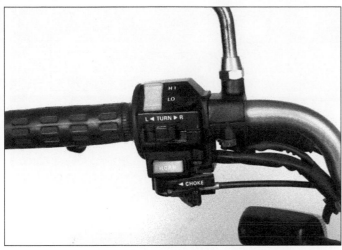

18.1a The left handlebar switches should look like this when assembled (YX600 Radian shown) . . .

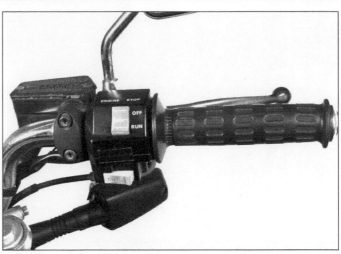

18.1b . . . and the right handlebar switches should look like this

18.1c Remove the screws from the bottom of the switch assembly and separate the halves (on the throttle side, one screw secures the throttle cable)

18.1d On YX600 Radian models, remove the choke lever and cable if you plan to replace the left handlebar switch

17 Handlebar switches - check

1 Generally speaking, the switches are reliable and trouble-free. Most troubles, when they do occur, are caused by dirty or corroded contacts, but wear and breakage of internal parts is a possibility that should not be overlooked. If breakage does occur, the entire switch and related wiring harness will have to be replaced with a new one, since individual parts are not usually available.

2 The switches can be checked for continuity with an ohmmeter or a continuity test light. Always disconnect the battery negative cable, which will prevent the possibility of a short circuit, before making the checks.

3 Trace the wiring harness of the switch in question and unplug the electrical connectors.

4 Find the continuity diagram for the switch being checked in the appropriate wiring diagram at the end of this Chapter.

5 Using the ohmmeter or test light, check for continuity between the terminals of the switch harness with the switch in the various positions. Continuity should exist between the terminals connected by a solid line when the switch is in the indicated position. For example, in illustration 16.2, the solid line between the red and blue terminals

indicates that there should be continuity between these two terminals when the key (and therefore the switch) is in the P position.

6 If the continuity check indicates a problem exists, refer to Section 18, remove the switch and spray the switch contacts with electrical contact cleaner. If they are accessible, the contacts can be scraped clean with a knife or polished with crocus cloth. If switch components are damaged or broken, it will be obvious when the switch is disassembled.

18 Handlebar switches - removal and installation

Refer to illustrations 18.1a through 18.1d

1 The handlebar switches are composed of two halves that clamp around the bars (see illustrations). They are easily removed for cleaning or inspection by taking out the clamp screws and pulling the switch halves away from the handlebars (see illustrations).

2 To completely remove the switches, the electrical connectors in the wiring harness should be unplugged.

3 When installing the switches, make sure the wiring harnesses are properly routed to avoid pinching or stretching the wires.

19.6 Remove the small screw and disconnect the wires, then remove three larger screws and remove the switch

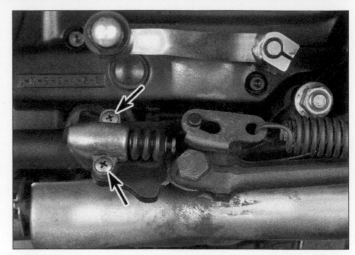

20.4 The sidestand switch is secured by two screws (arrows)

19 Neutral switch - check and replacement

Check

1   Make sure the transmission is in neutral.
2   Follow the switch harness (it comes from behind the engine sprocket cover on the left side of the engine) to its connector, then unplug the connector.
3   Locate the sky blue wire's terminal in the harness side of the connector (not the side of the connector that goes to the neutral switch). Connect the terminal to ground (earth) (bare metal on the motorcycle frame) with a short length of wire. Turn the ignition switch On.
   a) If the light stays out, check the bulb and the wiring between the ignition switch and neutral switch.
   b) If the neutral indicator light comes on, the neutral switch may be bad. Connect an ohmmeter between the sky blue terminal in the switch side of the connector and ground (earth). Shift through the gears. The ohmmeter should indicate continuity in neutral and infinite resistance in all other gears. If not, replace the neutral switch.

Replacement

Refer to illustration 19.6

4   Remove the engine sprocket cover (see Chapter 6).
5   Unplug the electrical connector (if you haven't already done so). Unbolt the wiring harness retainers from the crankcase and oil pan.
6   Loosen the screw and disconnect the wire from the switch **(see illustration)**. Remove the switch mounting screws and detach the switch from the crankcase.
7   Installation is the reverse of the removal steps. Be sure to install a new O-ring and/or gasket when you reinstall the switch.

20 Sidestand switch - check and replacement

Check

1   Follow the wiring harness from the switch to the connector, then unplug the connector. Connect the leads of an ohmmeter to the wire terminals. With the sidestand in the up position, there should be continuity through the switch (0 ohms).
2   With the sidestand in the down position, the meter should indicate infinite resistance.
3   If the switch fails either of these tests, replace it.

Replacement

Refer to illustration 20.4

4   With the sidestand in the up position, unscrew the two screws and remove the switch **(see illustration)**. Disconnect the switch electrical connector.
5   Installation is the reverse of the removal procedure.

21 Clutch switch - check and replacement

Refer to illustration 21.1

Check

1   Disconnect the electrical connector from the clutch switch **(see illustration)**.
2   Connect an ohmmeter between the terminals in the clutch switch. With the clutch lever pulled in, the ohmmeter should show continuity (little or no resistance). With the lever out, the ohmmeter should show infinite resistance.
3   If the switch doesn't check out as described, replace it.

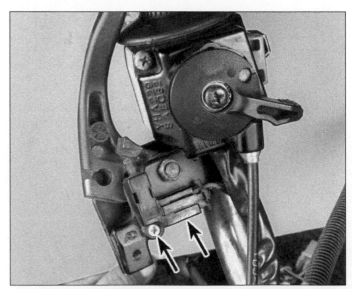

21.1 Disconnect the electrical connector and remove the screw (arrows) to remove the clutch switch

22.1 Disconnect the wires and remove the bracket bolt to detach the horn

23.3 Unplug the wiring connector, then remove the nuts (arrows) and disconnect the cables . . .

Replacement

4   If you haven't already done so, unplug the wiring connector. Remove the mounting screw and take the switch off **(see illustration 21.1)**.
5   Installation is the reverse of removal.

22 Horn - check and replacement

Check

Refer to illustration 22.1
1   Unplug the electrical connectors from the horn **(see illustration)**. Using two jumper wires, apply battery voltage directly to the terminals on the horn. If the horn sounds, check the switch (see Section 18) and the wiring between the switch and the horn (see the *wiring diagrams* at the end of this Chapter).
2   If the horn doesn't sound, replace it.

Replacement

3   Unbolt the horn bracket from the lower triple clamp **(see illustration 22.1)** and detach the electrical connectors.
4   Unbolt the horn from the bracket and transfer the bracket to the new horn.
5   Installation is the reverse of removal.

23 Starter relay - check and replacement

Check

Refer to illustration 23.3
1   Remove the seat (see Chapter 7) and the fuel tank (see Chapter 3). Remove the side cover on the left side of the motorcycle (see Chapter 7).
2   Disconnect the starter cable from the starter motor **(see illustration 24.6)**. Raise the sidestand. With the ignition switch On, the engine kill switch in Run and the transmission in neutral, press the starter switch. The relay should click.
3   If the relay doesn't click, disconnect the battery negative cable from the battery, then disconnect the positive cable from the battery and the relay **(see illustration)**. Disconnect the thin wire and remove the relay. Connect one lead of an ohmmeter to the terminal stud that the battery positive cable was connected to. Connect the other lead to the terminal on the thin wire. If the reading isn't within the range listed in this chapter's Specifications, replace the relay.

23.5 . . . and slide the starter relay (arrow) off its mounting tabs

Replacement

Refer to illustration 23.5
4   Disconnect the cable from the negative terminal of the battery.
5   Detach the battery positive cable, the starter cable and electrical connector from the relay **(see illustration)**.
6   Slide the relay off its mounting tabs.
7   Remove the relay from its rubber mounting.
8   Installation is the reverse of removal. Reconnect the negative battery cable after all the other electrical connections are made.

24 Starter motor - removal and installation

Removal

Refer to illustrations 24.6, 24.7 and 24.8

FZ600 models

1   Remove the center and lower fairing panels, the seats and the side covers (see Chapter 7).
2   Remove the air cleaner air box mounting bolts. Loosen the clamp screws that attach the carburetor air hoses to the air box and slide it backwards.

24.6   Pull back the rubber cover, remove the nut (arrow) and disconnect the starter cable

24.7   Remove the starter mounting bolts (arrows) . . .

24.8   . . . and pull the starter clear of the engine (engine removed for clarity)

25.2a   Remove the two long screws (the screw heads on some models are on the opposite end of the starter) . . .

YX600 Radian models

3   Remove the left footpeg bracket.

All models

4   Remove the shift pedal (see Chapter 2) and the engine sprocket cover (see Chapter 5).

5   Disconnect the cable from the negative terminal of the battery.

6   Remove the nut retaining the starter cable to the starter **(see illustration)**.

7   Remove the starter mounting bolts **(see illustration)**.

8   Pull the starter up slightly and slide the starter out of the engine case **(see illustration)**.

9   Check the condition of the O-ring on the end of the starter and replace it if necessary.

Installation

10   Apply a little engine oil to the O-ring and install the starter by reversing the removal procedure. Tighten the two mounting bolts to the torque listed in this chapter's Specifications.

---

25   Starter motor - disassembly, inspection and reassembly

---

1   Remove the starter motor (see Section 24).

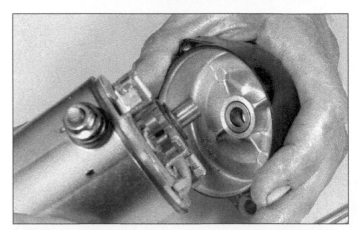

25.2b   . . . and remove the cover from each end of the starter

Disassembly

Refer to illustrations 25.2a, 25.2b, 25.2c, 25.3, 25.4 and 25.5

2   Mark the position of the housing to each end cover. Remove the two long screws and detach both end covers **(see illustrations)**.

3   Pull the armature out of the housing (toward the pinion gear side) **(see illustration)**.

4   Remove the brush plate from the housing **(see illustration)**.

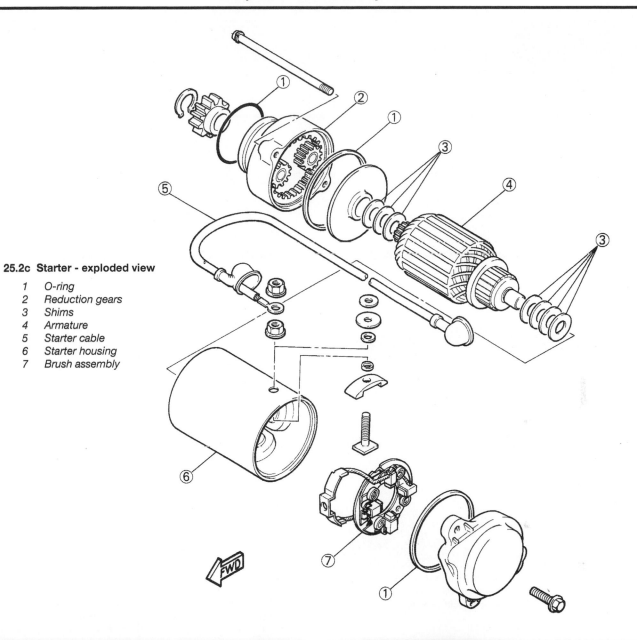

25.2c Starter - exploded view

1   O-ring
2   Reduction gears
3   Shims
4   Armature
5   Starter cable
6   Starter housing
7   Brush assembly

25.3 Lower the armature (arrow) out of the brush plate and starter housing

25.4 Remove the brush plate from the housing

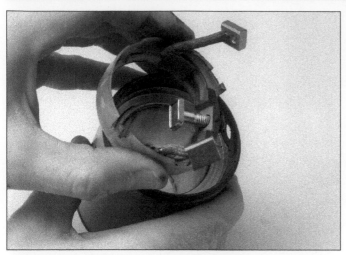

25.5 Push the terminal bolt through the housing and remove the brush plate

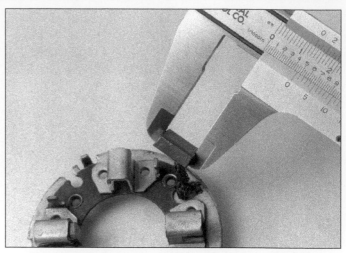

25.6 Measure the length of the brushes and compare the length of the shortest brush with the length listed in this Chapter's Specifications

5   Carefully note how the washers are arranged on the terminal bolt. Remove the nut and push the terminal bolt through the starter housing, then reinstall the washers and nut on the bolt so you don't forget how they go. Remove the two brushes with the plastic holder from the housing **(see illustration)**.

Inspection

Refer to illustrations 25.6, 25.7, 25.8a, 25.8b, 25.9 and 25.10

6   The parts of the starter motor that most likely will require attention are the brushes. Measure the length of the brushes and compare the results to the brush length listed in this chapter's Specifications **(see illustration)**. If any of the brushes are worn beyond the specified limits, replace the brush holder assembly with a new one. If the brushes are not worn excessively, nor cracked, chipped, or otherwise damaged, they may be reused.

7   Inspect the commutator **(see illustration)** for scoring, scratches and discoloration. The commutator can be cleaned and polished with crocus cloth, but do not use sandpaper or emery paper. After cleaning, wipe away any residue with a cloth soaked in an electrical system cleaner or denatured alcohol. Measure the commutator diameter and compare it to the diameter listed in this chapter's Specifications. If it is less than the service limit, the motor must be replaced with a new one.

8   Using an ohmmeter or a continuity test light, check for continuity between the commutator bars **(see illustration)**. Continuity should exist

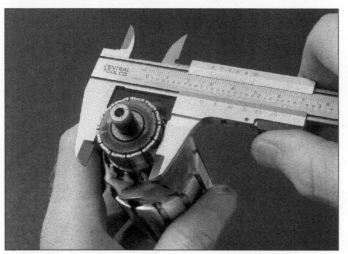

25.7 Check the commutator for cracks and discoloring, then measure the diameter and compare it with the minimum diameter listed in this Chapter's Specifications

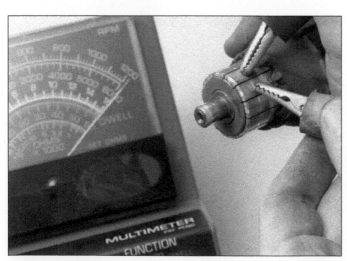

25.8a Continuity should exist between the commutator bars

25.8b There should be no continuity between the commutator bars and the armature shaft

25.9 There should be almost no resistance (0 ohms) between the brushes and the brush plate

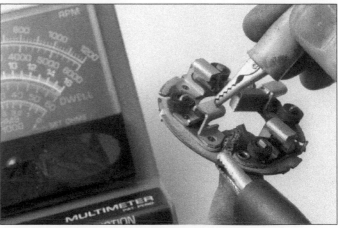

25.10 There should be no continuity between the brush plate and the brush holders (the resistance reading should be infinite)

25.13 When installing the brush plate, make sure the brush leads fit into the notches in the plate (arrow) - also, make sure the tongue on the plate fits into the notch in the housing (arrows)

25.14a Install each brush spring on the post in this position . . .

between each bar and all of the others. Also, check for continuity between the commutator bars and the armature shaft **(see illustration)**. There should be no continuity between the commutator and the shaft. If the checks indicate otherwise, the armature is defective.

9   Check for continuity between the brush plate and the brushes **(see illustration)**. The meter should read close to 0 ohms. If it doesn't, the brush plate has an open and must be replaced.

10   Using the highest range on the ohmmeter, measure the resistance between the brush holders and the brush plate **(see illustration)**. The reading should be infinite. If there is any reading at all, replace the brush plate.

11   Check the starter pinion gear for worn, cracked, chipped and broken teeth. If the gear is damaged or worn, replace the starter motor.

Reassembly

Refer to illustrations 25.13, 25.14a and 25.14b

12   Install the plastic brush holder into the housing. Make sure the terminal bolt and washers are assembled in their original order. Tighten the terminal nut securely.

13   Detach the brush springs from the brush plate (this will make armature installation much easier). Install the brush plate into the housing, routing the brush leads into the notches in the plate **(see illustration)**. Make sure the tongue on the brush plate fits into the notch in the housing.

14   Install the brushes into their holders and slide the armature into place. Install the brush springs **(see illustrations)**.

25.14b . . . then pull the end of the spring 1/2 turn counterclockwise (anti-clockwise) and seat the end of it in the groove in the end of the brush

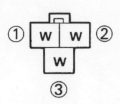

28.2 Connect the ohmmeter between all three pairs of terminals (connect it between terminal 1 and terminals 2 and 3 in turn, then between terminals 2 and 3); compare the resistance reading to this Chapter's Specifications

28.4 Connect the ohmmeter between the terminals in the field coil connector and compare the resistance reading to the Specifications

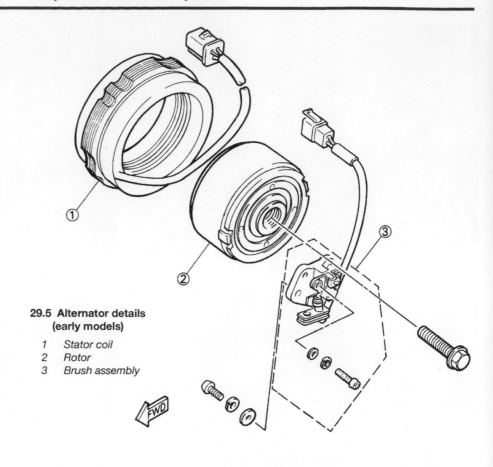

29.5 Alternator details (early models)

1   Stator coil
2   Rotor
3   Brush assembly

15   Install any washers that were present on the end of the armature shaft. Install the end covers, aligning the previously applied matchmarks. Install the two long screws and tighten them securely.

26   Charging system testing - general information and precautions

1   If the performance of the charging system is suspect, the system as a whole should be checked first, followed by testing of the individual components (the alternator and the voltage regulator/rectifier). **Note**: *Before beginning the checks, make sure the battery is fully charged and that all system connections are clean and tight.*

2   Checking the output of the charging system and the performance of the various components within the charging system requires the use of special electrical test equipment. A voltmeter or a multimeter are the absolute minimum tools required. In addition, an ohmmeter is required for checking the remainder of the system.

3   When making the checks, follow the procedures carefully to prevent incorrect connections or short circuits, as irreparable damage to electrical system components may result if short circuits occur. Because of the special tools and expertise required, it is recommended that the job of checking the charging system be left to a dealer service department or a reputable motorcycle repair shop.

27   Charging system - output test

Caution: *Never disconnect the battery cables from the battery while the engine is running. If the battery is disconnected, the alternator and regulator/rectifier will be damaged.*

1   To check the charging system output, you will need a voltmeter or a multimeter with a voltmeter function.

2   The battery must be fully charged (charge it from an external source if necessary) and the engine must be at normal operating temperature to obtain an accurate reading.

3   Attach the positive (red) voltmeter lead to the positive (+) battery terminal and the negative (black) lead to the battery negative (-) terminal. The voltmeter selector switch (if equipped) must be in a DC volt range greater than 15 volts.

4   Start the engine.

5   The charging system output should be within the range listed in this chapter's Specifications.

6   If the output is as specified, the alternator is functioning properly. If the charging system as a whole is not performing as it should, refer to Section 28 and test the system further.

7   Low voltage output may be the result of damaged windings in the alternator stator coils or wiring problems. Make sure all electrical connections are clean and tight, then refer to Sections 28 and 29 for brush replacement (early models only) and further tests.

8   High voltage output (above the specified range) indicates a defective voltage regulator/rectifier.

28   Charging system - component tests

Refer to illustrations 28.2 and 28.4

1   Follow the stator coil wiring harness (three white wires) from the alternator on the left side of the engine to the connector and unplug the connector.

2   Connect an ohmmeter to all three pairs of terminals in the connector (the alternator side of the connector, not the wiring harness side) and compare the resistance readings to those listed in this

29.9 On later models, remove the alternator cover together with the stator coil and pickup coil; note the cover dowel locations

29.10a Loosen the rotor bolt . . .

Chapter's Specifications **(see illustration)**. If the readings are not within specifications, the stator coil is defective. Replace it.

3   On early models (equipped with alternator brushes), follow the field coil harness (green and brown wires) from the alternator to the connector and unplug the connector.

4   Connect an ohmmeter between the terminals in the alternator side of the connector and compare the resistance reading to that listed in this Chapter's Specifications **(see illustration)**. If it's incorrect, check the brushes for wear and replace as necessary (see Section 29).

29   Alternator - removal and installation

1   Disconnect the cable from the negative terminal of the battery.
2   Remove body panels as necessary for access (see Chapter 7).

Early models (with brushes)

Refer to illustration 29.5

3   *Remove three Allen bolts and detach the alternator cover from the engine (it's on the left side behind the signal generator cover).*

4   Unplug the alternator electrical connectors.

5   Remove the brush assembly **(see illustration)**.
6   Remove the stator coil.
7   Place the transmission in gear and have an assistant hold the rear brake on to keep the engine from turning. Remove the rotor bolt and take off the rotor.
8   Installation is the reverse of the removal steps. Tighten the rotor bolt and brush assembly screws to the torques listed in this Chapter's Specifications.

Later models (without brushes)

Refer to illustrations 29.9, 29.10a, 20.10b, 29.11a, 29.11b, 29.11c, 29.12, 29.13a and 29.13b

9   Remove the alternator wiring harness retainers on the left side of the engine. Remove the alternator cover bolts and remove the alternator cover, together with the stator coil and pickup coil **(see illustration)**.

10   Place the transmission in gear and have an assistant hold the rear brake on to keep the engine from turning. Loosen the rotor bolt, then remove the bolt and washer **(see illustrations)**.

11   Thread a rotor puller into the rotor **(see illustration)**. Remove the rotor from the end of the crankshaft and take the Woodruff key out of its slot **(see illustrations)**.

29.10b . . . unscrew the bolt and remove the washer

29.11a Use a tool like this one to separate the rotor from the crankshaft . . .

29.11b ... then take the rotor off ...

29.11c ...and lift the Woodruff key out of its slot

29.12 Remove the stator Allen bolts and the pickup coil screws

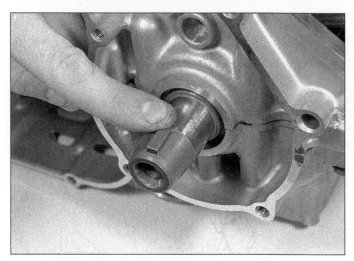

29.13a Seat the Woodruff key securely in its slot

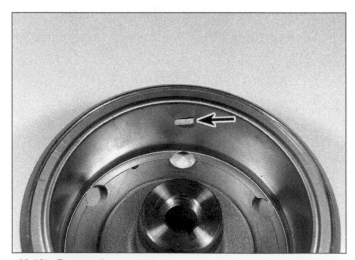

29.13b Be sure there aren't any metal objects stuck to the rotor magnets; an inconspicuous item like this Woodruff key (arrow) can ruin the rotor and stator if the engine is run

12   To remove the stator coil and pickup coil, remove the Allen bolts and Phillips screws and detach the coils from the alternator cover **(see illustration)**.

13   Installation is the reverse of the removal steps, with the following additions:

a) *Be sure to reinstall the Woodruff key and make sure no metal objects have stuck to the magnets inside the rotor* **(see illustrations)**.

b) *Tighten the rotor bolt, stator coil Allen bolts, pickup coil screws and wiring harness retainers to the torques listed in this Chapter's Specifications.*

c) *Use a new cover gasket and make sure the dowels are in position. Tighten the cover bolts to the torque listed in this Chapter's Specifications.*

30   Regulator/rectifier - removal and installation

Refer to illustration 30.2

1   The regulator is mounted on the left side of the motorcycle under the side cover (FJ/XJ600 and FZ600 models), on the air cleaner air box cover (early YX600 Radian models) or on the frame forward of the engine (later YX600 Radian models).

2   To remove the regulator/rectifier, remove body panels as necessary for access (see Chapter 7). Disconnect the electrical connector, remove the mounting screws and take the unit out **(see illustration)**.

3   Installation is the reverse of the removal steps.

31   Wiring diagrams

Prior to troubleshooting a circuit, check the fuses to make sure they're in good condition. Make sure the battery is fully charged and check the cable connections.

When checking a circuit, make sure all connectors are clean, with no broken or loose terminals or wires. When unplugging a connector, don't pull on the wires - pull only on the connector housings themselves.

Refer to the accompanying table for the wire color codes.

30.2 Unplug the connector and remove the mounting screws to remove the regulator/rectifier (later YX600 Radian shown)

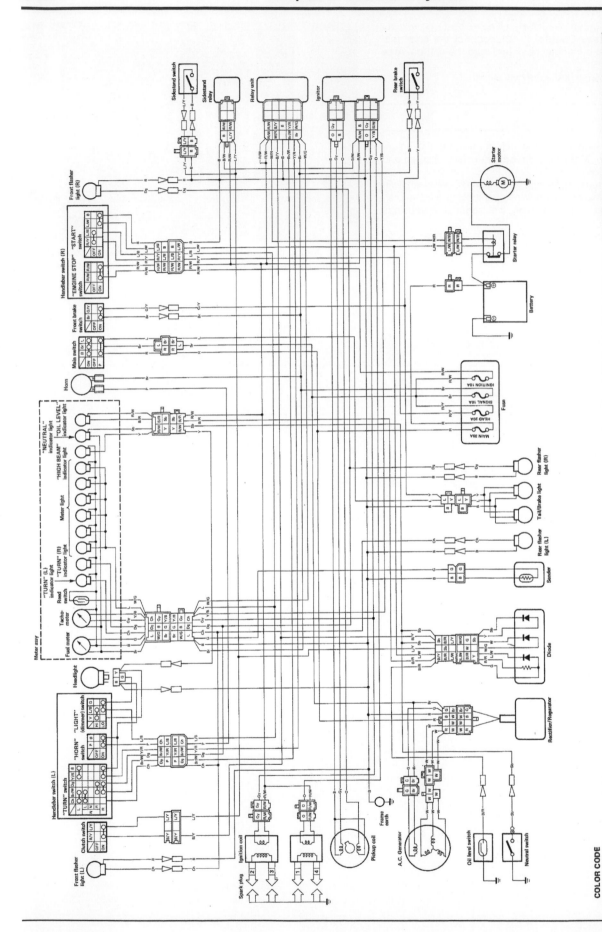

Wiring diagram - FJ600 models

COLOR CODE

B	Black	P	Pink	O	Orange	Br	Brown	Gy	Gray	L/Y	Blue/Yellow	B/W	Black/White	Y/B	Yellow/Black	W/R	White/Red
R	Red	G	Green	Y	Yellow	Sb	Sky blue	R/W	Red/White	L/W	Blue/White	B/Y	Black/Yellow	Y/R	Yellow/Red	Br/W	Brown/White
L	Blue	W	White	Dg	Dark green	Ch	Chocolate	R/Y	Red/Yellow	L/B	Blue/Black	B/R	Black/Red	W/G	White/Green	G/Y	Green/Yellow

Wiring diagram (US FZ600 models)

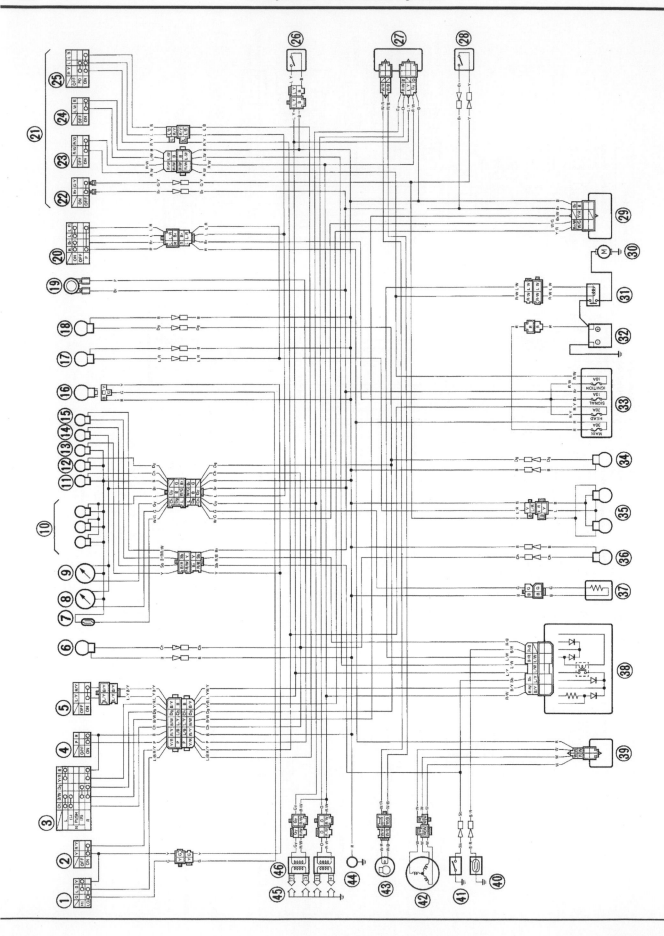

Wiring diagram (XJ600 models)

Wiring diagram key - UK FZ600 models (diagram on following page)

1 Pass switch
2 Clutch switch
3 Turn signal switch
4 Horn switch
5 Dimmer switch
6 Left front turn signal
7 High beam headlight
8 Speedometer
9 Fuel gauge
10 Tachometer
11 Instrument lights
12 Turn indicator light
13 High beam indicator light
14 Neutral indicator light
15 Oil level indicator light
16 Low beam headlight
17 Auxiliary light
18 Right front turn signal
19 Horn
20 Ignition main (key) switch
21 Front brake switch
22 Engine kill switch
23 Starter switch
24 Lighting switch
25 Igniter unit
26 Rear brake switch
27 Flasher relay
28 Starter motor
29 Starter relay
30 Battery
31 Fuse block
32 Right rear turn signal
33 Tail/brake light
34 Left rear turn signal
35 Fuel sender
36 Diode assembly
37 Rectifier/regulator
38 Neutral switch
39 Oil level switch
40 Alternator
41 Pickup coil
42 Body ground
43 Spark plugs
44 Ignition coils

COLOR CODE

B	Black	P	Pink	O	Orange	Br	Brown	Gy	Gray	L/Y	Blue/Yellow	B/W	Black/White	Y/B	Yellow/Black	W/R	White/Red
R	Red	G	Green	Y	Yellow	Sb	Sky blue	R/W	Red/White	L/W	Blue/White	B/Y	Black/Yellow	Y/R	Yellow/Red	Br/W	Brown/White
L	Blue	W	White	Dg	Dark green	Ch	Chcolate	R/Y	Red/Yellow	L/B	Blue/Black	B/R	Black/Red	W/G	White/Green	G/Y	Green/Yellow

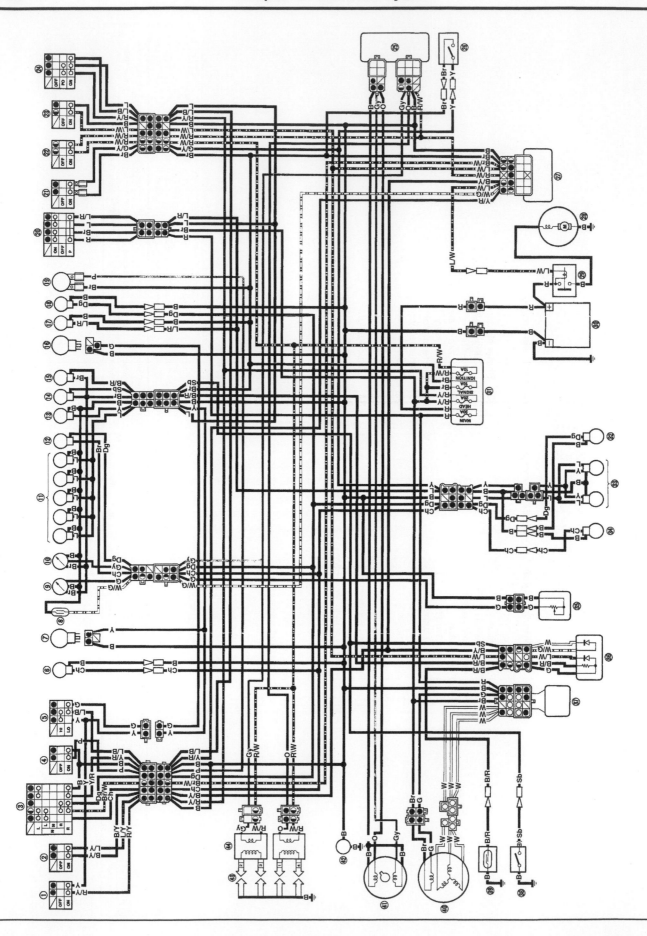

① Front flasher light (L)
② Clutch switch
③ "TURN" switch
④ "HORN" switch
⑤ "LIGHTS" (dimmer) switch
⑥ Headlight
⑦ "HIGH BEAM" indicator light
⑧ "NEUTRAL" indicator light
⑨ "OIL" indicator light
⑩ "TURN" indicator light
⑪ Reed switch
⑫ Meter light
⑬ Tachometer
⑭ Horn
⑮ Main switch
⑯ Front brake switch
⑰ "START" switch
⑱ "ENGINE STOP" switch
⑲ Front flasher light (R)
⑳ Sidestand relay
㉑ Sidestand switch
㉒ Relay assembly
㉓ Ignitor unit
㉔ Rear brake switch
㉕ Starter motor
㉖ Starter relay
㉗ Battery
㉘ Fuse
㉙ License light
㉚ Rear flasher light (R)
㉛ Tail/Brake light
㉜ Rear flasher light (L)
㉝ Air vent control valve (For YX600SC only)
㉞ Diode
㉟ Rectifier/regulator
㊱ Neutral switch
㊲ Oil level switch
㊳ AC magneto
㊵ Body earth
㊶ Pickup coil
㊷ Spark plug
㊸ Ignition coil

R/W . . . . Red/White
R/Y . . . . Red/Yellow
B/W . . . . Black/White
B/Y . . . . Black/Yellow
B/R . . . . Black/Red
W/R . . . . White/Red
W/G . . . . White/Green
Y/R . . . . Yellow/Red
G/Y . . . . Green/Yellow
L/W . . . . Blue/White
L/Y . . . . Blue/Yellow
L/B . . . . Blue/Black
Br/W . . . . Brown/White

COLOR CODE
R . . . . Red
B . . . . Black
W . . . . White
Y . . . . Yellow
G . . . . Green
L . . . . Blue
Br . . . . Brown
Ch . . . . Chocolate
Dg . . . . Dark green
P . . . . Pink
O . . . . Orange
Sb . . . . Sky blue
Gy . . . . Gray

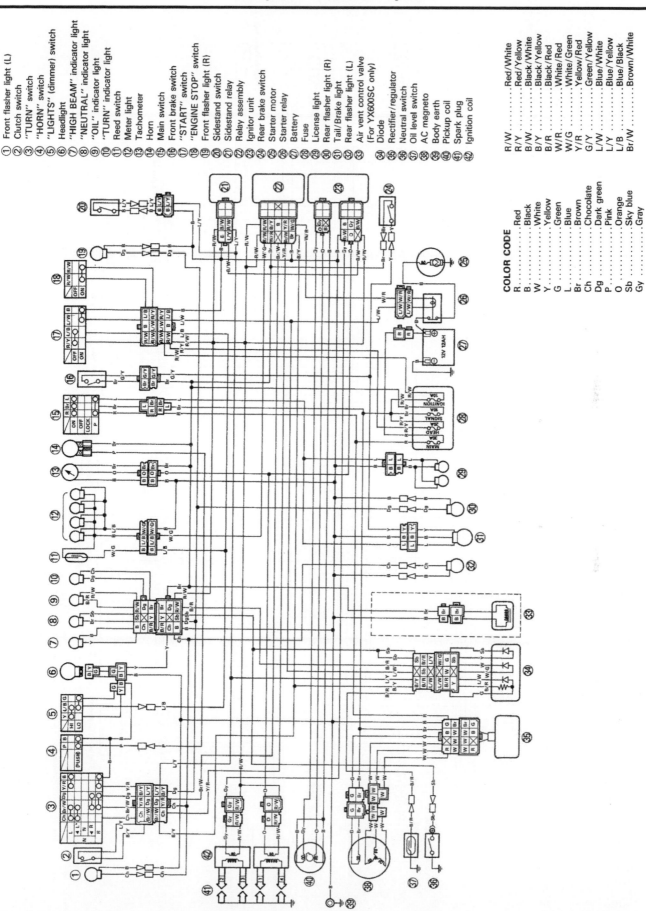

Wiring diagram (early YX600 Radian models)

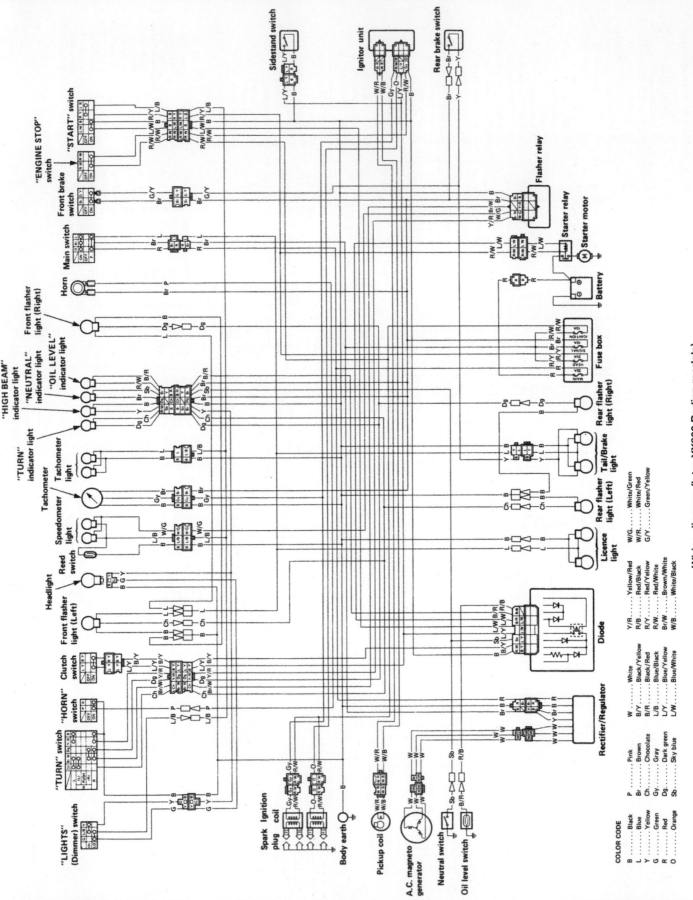

Wiring diagram (later YX600 Radian models)

Conversion factors

Length (distance)

Inches (in)	X	25.4	= Millimeters (mm)	X 0.0394	= Inches (in)
Feet (ft)	X	0.305	= Meters (m)	X 3.281	= Feet (ft)
Miles	X	1.609	= Kilometers (km)	X 0.621	= Miles

Volume (capacity)

Cubic inches (cu in; in³)	X	16.387	= Cubic centimeters (cc; cm³)	X 0.061	= Cubic inches (cu in; in³)
Imperial pints (Imp pt)	X	0.568	= Liters (l)	X 1.76	= Imperial pints (Imp pt)
Imperial quarts (Imp qt)	X	1.137	= Liters (l)	X 0.88	= Imperial quarts (Imp qt)
Imperial quarts (Imp qt)	X	1.201	= US quarts (US qt)	X 0.833	= Imperial quarts (Imp qt)
US quarts (US qt)	X	0.946	= Liters (l)	X 1.057	= US quarts (US qt)
Imperial gallons (Imp gal)	X	4.546	= Liters (l)	X 0.22	= Imperial gallons (Imp gal)
Imperial gallons (Imp gal)	X	1.201	= US gallons (US gal)	X 0.833	= Imperial gallons (Imp gal)
US gallons (US gal)	X	3.785	= Liters (l)	X 0.264	= US gallons (US gal)

Mass (weight)

Ounces (oz)	X	28.35	= Grams (g)	X 0.035	= Ounces (oz)
Pounds (lb)	X	0.454	= Kilograms (kg)	X 2.205	= Pounds (lb)

Force

Ounces-force (ozf; oz)	X	0.278	= Newtons (N)	X 3.6	= Ounces-force (ozf; oz)
Pounds-force (lbf; lb)	X	4.448	= Newtons (N)	X 0.225	= Pounds-force (lbf; lb)
Newtons (N)	X	0.1	= Kilograms-force (kgf; kg)	X 9.81	= Newtons (N)

Pressure

Pounds-force per square inch (psi; lbf/in²; lb/in²)	X	0.070	= Kilograms-force per square centimeter (kgf/cm²; kg/cm²)	X 14.223	= Pounds-force per square inch (psi; lbf/in²; lb/in²)
Pounds-force per square inch (psi; lbf/in²; lb/in²)	X	0.068	= Atmospheres (atm)	X 14.696	= Pounds-force per square inch (psi; lbf/in²; lb/in²)
Pounds-force per square inch (psi; lbf/in²; lb/in²)	X	0.069	= Bars	X 14.5	= Pounds-force per square inch (psi; lbf/in²; lb/in²)
Pounds-force per square inch (psi; lbf/in²; lb/in²)	X	6.895	= Kilopascals (kPa)	X 0.145	= Pounds-force per square inch (psi; lbf/in²; lb/in²)
Kilopascals (kPa)	X	0.01	= Kilograms-force per square centimeter (kgf/cm²; kg/cm²)	X 98.1	= Kilopascals (kPa)

Torque (moment of force)

Pounds-force inches (lbf in; lb in)	X	1.152	= Kilograms-force centimeter (kgf cm; kg cm)	X 0.868	= Pounds-force inches (lbf in; lb in)
Pounds-force inches (lbf in; lb in)	X	0.113	= Newton meters (Nm)	X 8.85	= Pounds-force inches (lbf in; lb in)
Pounds-force inches (lbf in; lb in)	X	0.083	= Pounds-force feet (lbf ft; lb ft)	X 12	= Pounds-force inches (lbf in; lb in)
Pounds-force feet (lbf ft; lb ft)	X	0.138	= Kilograms-force meters (kgf m; kg m)	X 7.233	= Pounds-force feet (lbf ft; lb ft)
Pounds-force feet (lbf ft; lb ft)	X	1.356	= Newton meters (Nm)	X 0.738	= Pounds-force feet (lbf ft; lb ft)
Newton meters (Nm)	X	0.102	= Kilograms-force meters (kgf m; kg m)	X 9.804	= Newton meters (Nm)

Vacuum

Inches mercury (in. Hg)	X	3.377	= Kilopascals (kPa)	X 0.2961	= Inches mercury
Inches mercury (in. Hg)	X	25.4	= Millimeters mercury (mm Hg)	X 0.0394	= Inches mercury

Power

Horsepower (hp)	X	745.7	= Watts (W)	X 0.0013	= Horsepower (hp)

Velocity (speed)

Miles per hour (miles/hr; mph)	X	1.609	= Kilometers per hour (km/hr; kph)	X 0.621	= Miles per hour (miles/hr; mph)

Fuel consumption*

Miles per gallon, Imperial (mpg)	X	0.354	= Kilometers per liter (km/l)	X 2.825	= Miles per gallon, Imperial (mpg)
Miles per gallon, US (mpg)	X	0.425	= Kilometers per liter (km/l)	X 2.352	= Miles per gallon, US (mpg)

Temperature

Degrees Fahrenheit = (°C x 1.8) + 32

Degrees Celsius (Degrees Centigrade; °C) = (°F - 32) x 0.56

It is common practice to convert from miles per gallon (mpg) to liters/100 kilometers (l/100km), where mpg (Imperial) x l/100 km = 282 and mpg (US) x l/100 km = 235

Index

Notes

Service record

Date	Mileage/hours	Work performed

Service record

Date	Mileage/hours	Work performed